Chapter and Unit Tests with Answer Keys

D1507454

HOLT, RINEHART AND **WINSTON**

Harcourt Brace & Company

Austin • New York • Orlando • Atlanta • San Francisco • Boston • Dallas • Toronto • London

ISBN 0-03-051442-8   5   021   00   99

Contents

About CHAPTER AND UNIT TESTS

The CHAPTER AND UNIT TESTS for *Modern Earth Science* are presented
as blackline masters. Each test can be reproduced and distributed to students.
The *Chapter and Unit Tests* booklet includes the following:

- 30 chapter tests
- 8 unit tests
- 1 comprehensive final exam
- answer keys

Each question in *Chapter and Unit Tests* corresponds directly to one of the
section objectives for a chapter. A comprehensive list of the objectives
found in *Modern Earth Science* is provided on pages 3–10. The section
objectives for each chapter are numbered sequentially. A number in
parentheses is found after each section objective. For each test question there
is a number in the answer keys corresponding to one of the section
objectives for a chapter.

Chapter Tests

A chapter test is provided for each chapter in *Modern Earth Science*.
Each chapter test is composed of 35 questions. The questions on each
test may include true-false questions, fill-in-the-blanks, multiple-choice
questions, and labeling diagrams. Each chapter test concludes with a
battery of essay questions. The questions range in level of difficulty,
indicated by A and B. A-level questions are for all students; B-level
questions are for above average students. A and B designations can be
found in the answer keys.

Unit Tests

A five- or six-page test is provided for each of the eight units in *Modern
Earth Science*. Each unit test is composed of a variety of 50 questions.

Final Examination

The comprehensive final examination is composed of 100 questions. The
final examination includes a variety of questions testing each chapter in
Modern Earth Science.

Section Objectives

The following is a list of the objectives found in *Modern Earth Science*. The objectives are listed by chapter section and numbered sequentially in parentheses after each section objective. The objectives can be used as a study guide or as a tool for outlining each chapter and section. The questions in the *Modern Earth Science* CHAPTER AND UNIT TESTS are correlated to the section objectives.

Unit 1 Studying the Earth
Chapter 1—Introduction to Earth Science
1.1 What Is Earth Science?
Section Objectives
- Name the four main branches of earth science. (1)
- Discuss the relationship between earth science and ecology. (2)

1.2 Paths to Discovery: Scientific Methods
Section Objectives
- Identify the steps that make up scientific methods. (3)
- Explain how the meteorite-impact hypothesis developed. (4)

1.3 Birth of a Theory: The Big Bang
Section Objectives
- Distinguish between a hypothesis, a theory, and a scientific law. (5)
- Describe the Doppler effect. (6)
- Summarize the Big Bang theory of the origin of the universe. (7)
- List evidence for the Big Bang theory. (8)

Chapter 2—The Earth in Space
2.1 Earth: A Unique Planet
Section Objectives
- List the characteristics of the earth's three major zones. (1)
- Explain how studies of seismic waves have provided information about the earth's interior. (2)
- Define *magnetosphere* and identify the possible source of the earth's magnetism. (3)
- Summarize Newton's law of gravitation. (4)

2.2 Movements of the Earth
Section Objectives
- Describe the earth's revolution and rotation. (5)
- Tell why the seasons change. (6)
- Explain how the sun is used as a basis for measuring time. (7)

2.3 Artificial Satellites
Section Objectives
- Compare two types of satellite orbits. (8)
- Discuss ways in which satellites are used to study the earth. (9)

Chapter 3—Models of the Earth
3.1 Finding Locations on the Earth
Section Objectives
- Distinguish between latitude and longitude. (1)
- Explain how latitude and longitude can be used to locate places on the earth. (2)
- Explain how a magnetic compass can be used to find directions on the earth. (3)

3.2 Mapping the Earth's Surface
Section Objectives
- Describe the characteristics and uses of three types of map projections. (4)
- Define *scale* and explain how scale can be used to find distance on a map. (5)

3.3 Topographic Maps
Section Objectives
- Explain how elevation and topography can be shown on a map. (6)
- Interpret a topographic map. (7)

8.2 Combination of Atoms
Section Objectives
- Explain how atoms join together to form compounds. (6)
- Describe two ways that electrons form chemical bonds between atoms. (7)
- Read and interpret chemical formulas. (8)
- Explain the difference between compounds and mixtures. (9)

Chapter 9—Minerals of the Earth's Crust
9.1 What Is a Mineral?
Section Objectives
- Define a mineral and distinguish between the two main mineral groups. (1)
- Identify the elements found most abundantly in common minerals. (2)
- Name six types of nonsilicate minerals. (3)
- Distinguish among four main arrangements of silicon-oxygen tetrahedra found in silicate minerals. (4)

9.2 Identifying Minerals
Section Objectives
- Describe some characteristics that help distinguish one mineral from another. (5)
- List four special properties that may help identify certain minerals. (6)

Chapter 10—Rocks
10.1 Rocks and the Rock Cycle
Section Objectives
- Identify the three major types of rock, and explain how each is formed. (1)
- Summarize the steps in the rock cycle. (2)

10.2 Igneous Rock
Section Objectives
- Describe how the cooling rate of magma and lava affects the structure of igneous rocks. (3)
- Classify igneous rocks according to their mineral composition. (4)
- Describe a number of identifiable igneous rock structures. (5)

10.3 Sedimentary Rock
Section Objectives
- Name the three main types of sedimentary rock and give an example of each. (6)
- Describe several identifiable sedimentary rock features. (7)

10.4 Metamorphic Rock
Section Objectives
- Distinguish between regional and contact metamorphism. (8)
- Distinguish between foliated and unfoliated metamorphic rocks and give an example of each. (9)

Chapter 11—Resources and Energy
11.1 Mineral Resources
Section Objectives
- Explain what ores are and how they form. (1)
- Discuss the wide variety of uses for mineral resources. (2)

11.2 Fossil Fuels
Section Objectives
- Explain why coal is a fossil fuel. (3)
- Describe how petroleum and natural gas are formed and how they are removed from the earth. (4)
- Discuss the importance of fossil fuels as a source of energy and of petrochemical products. (5)
- Explain that fossil fuels are nonrenewable resources that must be used wisely. (6)
- Describe some of the effects that the use of fossil fuels has on the environment. (7)

11.3 Nuclear Energy
Section Objectives
- Summarize the process of nuclear fission. (8)
- Summarize the process of nuclear fusion. (9)

11.4 Alternate Energy Sources
Section Objectives
- Describe passive and active methods of harnessing energy from the sun. (10)
- Explain how geothermal energy may be a substitute for fossil fuels. (11)
- Discuss hydroelectric energy and wind energy as alternate energy sources. (12)

Unit 4 Reshaping the Crust
Chapter 12—Weathering and Erosion
12.1 Weathering Processes
Section Objectives
- Discuss the agents of mechanical weathering. (1)
- Discuss the chemical reactions that decompose rocks. (2)

12.2 Rates of Weathering

Section Objectives

- Explain how rock composition affects the rate of weathering. (3)
- Discuss how the amount of exposure determines the rate at which rock weathers. (4)
- Describe the effects of climate on the rate of weathering. (5)

12.3 Weathering and Soil

Section Objectives

- Explain how the composition of parent rock affects soil composition. (6)
- Predict the type of soil produced in various climates. (7)

12.4 Erosion

Section Objectives

- Define *erosion* and list four agents of erosion. (8)
- Identify various unwise farming methods that result in accelerated soil erosion and list four methods of soil conservation that prevent this damage. (9)
- Discuss how mass movements contribute to erosion. (10)
- Describe the major landforms. (11)

Chapter 13—Water and Erosion

13.1 The Water Cycle

Section Objectives

- Outline the stages of the water cycle. (1)
- Explain the components of a water budget. (2)
- List two approaches to water conservation. (3)

13.2 River Systems

Section Objectives

- Describe how a river develops. (4)
- Explain how a stream causes erosion. (5)
- Distinguish youthful, mature, and old river valleys. (6)

13.3 Stream Deposition

Section Objectives

- List two types of stream deposition and explain the differences between them. (7)
- Describe the change in a stream that causes flooding. (8)
- Identify direct and indirect methods of flood control. (9)

Chapter 14—Groundwater and Erosion

14.1 Water Beneath the Surface

Section Objectives

- Distinguish between porosity and permeability. (1)
- Identify the two moisture zones beneath the earth's surface. (2)
- Relate the contour of the water table to the contour of the land. (3)
- Describe how groundwater can be polluted. (4)

14.2 Wells and Springs

Section Objectives

- Describe artesian formation. (5)
- Describe two land features produced when groundwater is heated beneath the surface. (6)

14.3 Groundwater and Chemical Weathering

Section Objectives

- Explain how caverns and sinkholes form. (7)
- Identify the features of karst topography. (8)

Chapter 15—Glaciers and Erosion

15.1 Glaciers: Moving Ice

Section Objectives

- Describe how glaciers form. (1)
- Compare two main kinds of glaciers. (2)
- Explain two processes by which glaciers move. (3)

15.2 Landforms Created by Glaciers

Section Objectives

- Describe the landscape features that are produced by glacial erosion. (4)
- Name and describe five features formed by glacial deposition. (5)
- Explain how lakes are formed by glacial action. (6)

15.3 Ice Ages

Section Objectives

- Describe the climatic cycles that exist during an ice age. (7)
- Identify and summarize the theory that best accounts for the ice ages. (8)

Chapter 16 – Erosion by Wind and Waves

16.1 Wind Erosion
Section Objectives
- Describe two ways that the wind erodes the land. (1)
- Compare the two types of wind deposits. (2)

16.2 Wave Erosion
Section Objectives
- Compare the formation of six features produced by wave erosion. (3)
- Define a beach and discuss the way in which it is formed. (4)
- Describe the movement of sand along a shore and the features it produces. (5)

16.3 Coastal Erosion and Deposition
Section Objectives
- Explain how changes in sea level relative to the land affect coastlines. (6)
- Describe the features of a barrier island. (7)
- Compare the types of coral reef. (8)
- Analyze the effect of human activity on coastal land. (9)

Unit 5 The History of the Earth
Chapter 17—The Rock Record

17.1 Determining Relative Age
Section Objectives
- State the principle of uniformitarianism. (1)
- Explain how the law of superposition can be used to determine the relative age of rocks. (2)
- Compare three types of unconformity. (3)
- Apply the law of crosscutting relationships to determine the relative age of rocks. (4)

17.2 Determining Absolute Age
Section Objectives
- Summarize the limitations of using the rates of erosion and deposition to determine the absolute age of rocks. (5)
- Describe the formation of varves. (6)
- Explain how the process of radioactive decay can be used to determine the absolute age of rocks. (7)

17.3 The Fossil Record
Section Objectives
- Describe four ways in which entire organisms can be preserved as fossils. (8)
- List four examples of fossilized traces of organisms. (9)
- Describe how index fossils can be used to determine the relative age of rocks. (10)

Chapter 18—A View of the Earth's Past

18.1 The Geologic Time Scale
Section Objectives
- Summarize the development of the geologic column. (1)
- List the major units of geologic time. (2)

18.2 Geologic History
Section Objectives
- Identify the characteristics of Precambrian rock. (3)
- Explain what scientists have learned from the geologic record about life during the Paleozoic era. (4)
- Explain what scientists have learned from the geologic record about life during the Mesozoic era. (5)
- Explain what scientists have learned from the geologic record about life during the Cenozoic era. (6)

Chapter 19—The History of the Continents

19.1 Movements of the Continents
Section Objectives
- Identify the landmasses that made up Pangaea. (1)
- Describe the breakup of Pangaea. (2)

19.2 Growth of a Continent: North America
Section Objectives
- Summarize the changes in the North American continent that occurred during Precambrian time and the Paleozoic Era. (3)
- Summarize the changes in the North American continent that occurred during the Mesozoic and Cenozoic eras. (4)

19.3 Formation of the Grand Canyon
Section Objectives
- Interpret the rock record of the Grand Canyon, and state what it reveals about the geologic history of the canyon region. (5)
- Explain how fossils can reveal the history of the Grand Canyon. (6)

24.2 Clouds and Fog

Section Objectives
- List the conditions that must exist for a cloud to form. (4)
- Identify the types of clouds. (5)
- Describe four ways fog may form. (6)

24.3 Precipitation

Section Objectives
- Describe the various types of liquid and solid precipitation. (7)
- Compare the two processes that cause precipitation. (8)
- Describe how rain may be produced artificially. (9)
- Describe how precipitation is measured. (10)

Chapter 25—Weather

25.1 Air Masses

Section Objectives
- Explain how an air mass forms. (1)
- List and describe the types of air masses that usually affect the weather of North America. (2)

25.2 Fronts

Section Objectives
- Compare the characteristic weather patterns associated with cold fronts and warm fronts. (3)
- Describe how a wave cyclone forms. (4)
- Describe the stages in the development of hurricanes, thunderstorms, and tornadoes. (5)

25.3 Weather Instruments

Section Objectives
- Describe the types of instruments used to measure air temperature and wind speed. (6)
- Describe the instruments used to measure upper-atmospheric weather conditions. (7)

25.4 Forecasting the Weather

Section Objectives
- Explain how a weather map is made. (8)
- Describe the steps involved in preparing a weather forecast. (9)

Chapter 26—Climate

26.1 Factors That Affect Climate

Section Objectives
- Explain how latitude determines the amount of solar energy received on earth. (1)
- Describe how the different rates at which land and water are heated affect climate. (2)
- Explain the effects of topography on climate. (3)

26.2 Climate Zones

Section Objectives
- Name and describe the three types of tropical climates. (4)
- Compare subarctic and tundra climates. (5)
- List the various types of middle-latitude climates, and name the regions in which they are found. (6)
- Explain why city climates may differ from rural climates. (7)

Unit 8 Studying Space
Chapter 27—Stars and Galaxies

27.1 Characteristics of Stars

Section Objectives
- Describe how astronomers determine the composition and surface temperature of a star. (1)
- Explain why stars appear to move to an observer on the earth. (2)
- Name and describe the way astronomers measure the distance from the earth to the stars. (3)
- Explain the difference between absolute magnitude and apparent magnitude. (4)

27.2 Stellar Evolution

Section Objectives
- Describe how a protostar develops into a star. (5)
- Explain how a main-sequence star generates energy. (6)
- Describe the possible evolution of a star during and after the giant stage. (7)

27.3 Star Groups

Section Objectives
- Describe the characteristics that identify a constellation. (8)
- Describe the three main types of galaxies. (9)
- Explain the big bang theory. (10)

Chapter 28—The Sun

28.1 Structure of the Sun
Section Objectives
- Explain how the sun converts matter into energy in its core. (1)
- Compare the radiative and convective zones of the sun. (2)
- Describe the three layers of the sun's atmosphere. (3)

28.2 Solar Activity
Section Objectives
- Explain how sunspots are related to powerful magnetic fields on the sun. (4)
- Compare prominences and solar flares. (5)
- Describe how the solar wind can cause auroras on the earth. (6)

28.3 Formation of the Solar System
Section Objectives
- Explain the nebular theory of the origin of the solar system. (7)
- Describe how the planets developed. (8)
- Describe the formation of the land, the atmosphere, and the oceans of the earth. (9)

Chapter 29—The Solar System

29.1 Models of the Solar System
Section Objectives
- Compare the models of the universe developed by Ptolemy and Copernicus. (1)
- Summarize Kepler's three laws of planetary motion. (2)

29.2 The Inner Planets
Section Objectives
- Identify the basic characteristics of Mercury and Venus. (3)
- Identify the basic characteristics of Earth and Mars. (4)

29.3 The Outer Planets
Section Objectives
- Identify the basic characteristics of Jupiter and Saturn. (5)
- Identify the basic characteristics of Uranus, Neptune, and Pluto. (6)

29.4 Asteroids, Meteoroids, and Comets
Section Objectives
- Describe the physical characteristics of asteroids and of comets. (7)
- Compare and contrast meteoroids, meteorites, and meteors. (8)

Chapter 30—Moons and Rings

30.1 Earth's Moon
Section Objectives
- List the five kinds of lunar surface features. (1)
- Describe the interior of the moon. (2)
- Summarize the four stages in the development of the moon. (3)

30.2 Movements of the Moon
Section Objectives
- Describe the orbit of the moon around the earth. (4)
- Explain why eclipses occur. (5)

30.3 The Lunar Cycle
Section Objectives
- Describe the phases of the moon. (6)
- Explain how calendars are based on the movements of the earth and the moon. (7)

30.4 Satellites of Other Planets
Section Objectives
- Compare the characteristics of the two moons of Mars. (8)
- Compare the Galilean moons and the rings of Jupiter with the moons and rings of the other outer planets. (9)

M O D E R N   E A R T H   S C I E N C E

Earth Science
Safety Test

Read each statement below. If the statement is true, write *T* in the space provided. If the statement is false, write *F* in the space provided.

_____ **1.** The proper procedure for diluting an acid is to add water to the acid.

_____ **2.** Mouth-pipetting of chemicals is a dangerous practice.

_____ **3.** All electrical laboratory equipment should be properly grounded.

_____ **4.** It is safer to heat rocks directly over a flame than in a water bath.

_____ **5.** Pipettes and glass stirring rods should be dried by toweling.

Choose the one best response. Write the letter of that choice in the space provided.

_____ **6.** Which of the following symbols is used to indicate dangerous or caustic chemicals?

a.

b.

c.

d.

_____ **7.** Which of the following provides the best protection when working with toxic vapors?

    a. fire blanket   **b.** fume hood
    c. safety glasses   **d.** laboratory apron

MODERN EARTH SCIENCE

Safety Test

Choose the one best response. Write the letter of that choice in the space provided.

_____ **8.** Which of the following hazards does this symbol identify?

      a. fire   **b.** radiation
      c. bright light   **d.** explosion

_____ **9.** Which of the following symbols is used to indicate the need for special attention to hygiene?

   a.   **b.**

   c.   **d.**

_____ **10.** Which of the following safety guidelines is most closely associated with this symbol?

      a. Never look directly at the sun.
      b. Do not inhale fumes directly.
      c. Tie back long hair and confine loose clothing.
      d. Make sure an emergency eye wash station is available in the laboratory.

M O D E R N   E A R T H   S C I E N C E

Earth Science
Metric Test

**Read each statement below. If the statement is true, write *T* in the space provided.
If the statement is false, write *F* in the space provided.**

_____ **1.** In the metric system, the Fahrenheit degree is the basic unit of temperature.

_____ **2.** A meter is approximately equal to one foot.

_____ **3.** In the metric system, the prefix *kilo-* means 1,000.

_____ **4.** A temperature of 100°C is equal to a temperature of 32°F.

_____ **5.** In the metric system, the basic unit for measuring time is the second.

Choose the one best response. Write the letter of that choice in the space provided.

_____ **6.** Which of the following is equal to 500 m?

      a. 0.5 km   **b.** 5.0 km   **c.** 50 km   **d.** 5,000 km

_____ **7.** In order to determine the number of grams in 638 milligrams, you should:

      a. multiply by 100.   **b.** multiply by 1,000.
      c. divide by 100.   **d.** divide by 1,000.

_____ **8.** Which of the following is a metric unit of weight?

      a. ounce   **b.** gram   **c.** pound   **d.** newton

_____ **9.** How many centimeters are in one kilometer?

      a. 100   **b.** 1,000   **c.** 100,000   **d.** 1,000,000

_____ **10.** What is the volume of a box that has a length of 10 cm, a width of 5 cm, and a height of 40 cm?

      a. 1 liter   **b.** 2 liters   **c.** 10 liters   **d.** 20 liters

M O D E R N   E A R T H   S C I E N C E

Chapter 1
Introduction to Earth Science

**Read each statement below. If the statement is true, write *T* in the space provided.
If the statement is false, write *F* in the space provided.**

_____ **1.** Meteorology is the study of the earth's oceans.

_____ **2.** Most plastics are biodegradable.

_____ **3.** The physical environment of the earth supports the biosphere.

_____ **4.** A hypothesis may be based on facts established through experimentation.

_____ **5.** Scientists have simulated climatic conditions on a computer to help explain the extinction of the dinosaurs.

_____ **6.** According to the Big Bang theory, all galaxies in the universe are moving toward the earth.

_____ **7.** Scientists believe that an abnormally high amount of dust may have formed a cloud over the earth at one time.

_____ **8.** In order for a theory to become a law, it must be proven correct every time it is tested.

_____ **9.** As a light source moves away from an observer, the wavelengths of light will appear shorter to the observer.

_____ **10.** Earth science includes the study of oceans.

Choose the one best response. Write the letter of that choice in the space provided.

_____ **11.** Which of the following is an important means of gathering information?

     a. stating the problem   **b.** making a measurement
     c. forming a hypothesis   **d.** stating a conclusion

_____ **12.** Which of the following is a part of the biosphere?

     a. the oceans   **b.** the moon
     c. the earth's core   **d.** the sun

_____ **13.** Eratosthenes used careful observation and simple geometry to determine the:

     a. distance between the earth and the moon.
     b. circumference of the earth.
     c. distance between the earth and the sun.
     d. circumference of the sun.

M O D E R N · E A R T H   S C I E N C E

Chapter 1

Choose the one best response. Write the letter of that choice in the space provided.

_____ **14.** Which of the following is a meteorologist most likely to study?

 a. mountain formations   b. marine life
 c. rainfall trends   d. meteorites

_____ **15.** A spectroscope is most likely to be used to determine a star's:

 a. distance from earth.   b. diameter.
 c. chemical makeup.   d. brightness.

_____ **16.** A well-established and proven theory is most likely to become:

 a. a scientific law.   b. a hypothesis.
 c. an observation.   d. a conclusion.

_____ **17.** The discovery of background radiation evenly distributed throughout the universe provided support for the:

 a. Doppler effect.   b. meteorite-impact hypothesis.
 c. Big Bang theory.   d. law of gravitation.

_____ **18.** The meteorite-impact hypothesis is one explanation for the discovery of:

 a. background radiation in some galaxies.
 b. bright-line spectra in some elements.
 c. a red shift in some stars.
 d. high iridium levels in some rocks.

Use the diagram below to answer questions 19 and 20.

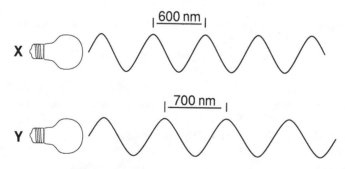

_____ **19.** Compared to the light coming from source **X**, the light coming from source **Y** appears:

 a. brighter.   b. less bright.   c. more red.   d. more blue.

_____ **20.** As an observer moves toward sources **X** and **Y**, light from the two sources will appear to become:

 a. red-shifted.   b. less bright.
 c. more red.   d. blue-shifted.

M O D E R N   E A R T H   S C I E N C E

Chapter 1

Complete each statement by writing the correct term or phrase in the space provided.

21. The field of science based on the study of the earth and of the universe around it is called

    _____ .

22. The term referring to the gases surrounding the earth is _____ .

23. The contamination of the environment with waste products is called

    _____ .

24. In the upper atmosphere, the form of oxygen that protects the earth from the sun's

    ultraviolet rays is called _____ .

25. To ensure that an experiment is testing only one variable, scientists often run a

    _____ .

26. The guidelines used by scientists to conduct their research are called

    _____ .

27. A scientist's possible explanation or solution to a problem is called a

    _____ .

28. Sunlight is made up of a mixture of different-colored light called the

    _____ .

29. A bright-line spectrum uniquely identifies each _____ .

30. Scientists determine the speed at which galaxies travel by examining the degree of red

    _____ .

Read each question or statement and answer it in the space provided.

31. Describe the effects of plastic waste on fish and other animals.

    _____

    _____

    _____

    _____

M O D E R N   E A R T H   S C I E N C E

Chapter 1

Read each question or statement and answer it in the space provided.

32. Describe the Big Bang hypothesis.

_____

_____

_____

_____

33. Describe the relationships between plants, animals, and microorganisms in a tropical rain forest.

_____

_____

_____

_____

34. Describe what changes in the spectra of galaxies tell scientists about the universe.

_____

_____

_____

_____

35. How might travelers in space use the Doppler effect to determine whether they are moving toward or away from a distant galaxy?

_____

_____

_____

_____

M O D E R N E A R T H S C I E N C E

Chapter 2

The Earth in Space

Read each statement below. If the statement is true, write *T* in the space provided. If the statement is false, write *F* in the space provided.

_____ **1.** Circadian rhythms occur in 24-hour cycles.

_____ **2.** The lithosphere is made up of the solid part of the upper mantle and the crust.

_____ **3.** Studies of seismic waves suggest that the lithosphere is probably more rigid than the asthenosphere.

_____ **4.** P waves travel faster through liquids than through solids.

_____ **5.** According to Newton's law of gravitation, the force of attraction between any two objects depends on the weight and shape of the objects.

_____ **6.** Rays of the sun that strike the earth at about 90° produce the greatest amount of heat on the earth's surface.

_____ **7.** When the Southern Hemisphere experiences summer, the Northern Hemisphere experiences winter.

_____ **8.** In a standard time zone, the sun is highest over the center of the zone at noon.

_____ **9.** A satellite in polar orbit passes over a different portion of the earth's surface during each orbit.

_____ **10.** The orbit of Landsat satellites around the earth is geosynchronous.

Choose the one best response. Write the letter of that choice in the space provided.

_____ **11.** Geologists believe the source of the earth's magnetic field may be in the:

    a. core.   **b.** magnetosphere.
    c. crust.   **d.** magnetic pole.

_____ **12.** What is the weight of a 5 kg mass on the earth's surface?

    a. 10 N   **b.** 50 N   **c.** 100 N   **d.** 500 N

_____ **13.** In what month does the earth reach perihelion?

    a. January   **b.** March   **c.** July   **d.** September

_____ **14.** The speed necessary to keep a satellite in orbit around the earth is determined primarily by the satellite's:

    a. orbit type.   **b.** altitude.   **c.** size.   **d.** weight.

M O D E R N   E A R T H   S C I E N C E

Chapter 2

Choose the one best response. Write the letter of that choice in the space provided.

_____ **15.** Which of the following is most likely to result from an increase in the earth's rotation rate?

        a. an increase in the length of a day
        b. an increase in the length of a year
        c. a decrease in the length of a day
        d. a decrease in the length of a year

_____ **16.** S waves are most easily blocked by the earth's:

        a. asthenosphere.   **b.** lithosphere.
        c. inner core.   **d.** outer core.

_____ **17.** Which of the following best describes the material that makes up the earth's asthenosphere?

        a. a rigid solid
        b. a solid that is able to flow
        c. a liquid at high temperature
        d. a gas under great pressure

_____ **18.** Landsat satellites are most frequently used to:

        a. identify features on the earth's surface.
        b. locate ships that are lost or in trouble.
        c. transmit communication signals across continents.
        d. analyze the atmosphere of planets.

Use the diagram below to answer questions 19 and 20.

TIME ZONES

_____ **19.** What is the correct time in Denver when it is noon in Boston?

    a. 9 A.M.   **b.** 10 A.M.   **c.** 1 P.M.   **d.** 2 P.M.

_____ **20.** The cities of St. Louis and Los Angeles are separated by approximately how many degrees of longitude?

    a. 2°   **b.** 15°   **c.** 30°   **d.** 90°

M O D E R N   E A R T H   S C I E N C E

Chapter 2

Complete each statement by writing the correct term or phrase in the space provided.

21. The sun's rays strike the earth at a 90° angle along the Tropic of Capricorn on the winter

_____ .

22. The boundary between the earth's crust and mantle is called the _____ .

23. Together, the earth's crust and upper mantle are called the _____ .

24. The force exerted on an object by the pull of gravity is called the object's

_____ .

25. The point at which a planet is closest to the sun is called _____ .

26. The wobbling of the earth's axis as it turns in space is called _____ .

27. The system under which clocks are set one hour ahead in April is called

_____ .

28. Jet lag is caused by an upset in the body's biological _____ .

29. A satellite that always remains above the same point on the equator is in an orbit called

_____ .

30. The point in a satellite's orbit that is closest to the earth is called

_____ .

Read each statement and answer it in the space provided.

31. Explain the relationship between weight and location on the earth's surface.

_____

_____

_____

_____

M O D E R N   E A R T H   S C I E N C E

Chapter 2

Read each statement and answer it in the space provided.

32. Describe how day length is affected by the tilt of the earth's axis.

_____

_____

_____

_____

33. Describe how changes in the angle of the sun's rays on the earth cause seasons in the Northern Hemisphere.

_____

_____

_____

_____

34. Describe Newton's law of gravitation.

_____

_____

_____

_____

35. When it is 4:00 P.M. on Thursday in Sydney, Australia, it is 6:00 P.M. on Wednesday in Nome, Alaska. Describe the relative positions of these cities in terms of time zones and the international date line.

_____

_____

_____

_____

Chapter 3
Models of the Earth

Read each statement below. If the statement is true, write T in the space provided. If the statement is false, write F in the space provided.

_____ 1. Meridians are circles around the earth parallel to the equator.

_____ 2. Each degree of latitude is equal to 60 seconds of latitude.

_____ 3. The earth's axis of rotation intersects the earth's surface at the geographic poles.

_____ 4. On a gnomonic projection, lines of longitude are parallel.

_____ 5. Landsat satellites are used by cartographers.

_____ 6. On a topographic map, elevation is measured as the distance above or below mean sea level.

_____ 7. On a topographic map, elevation is represented by contour lines.

_____ 8. True north is another name for the geomagnetic North Pole.

_____ 9. Most maps are drawn with south at the top.

_____ 10. The distance represented by a degree of longitude is greatest at the equator.

Choose the one best response. Write the letter of that choice in the space provided.

_____ 11. The North Pole has a latitude of:

      a. 0° N.   **b.** 45° N.   **c.** 90° N.   **d.** 180° N.

Use the information below to answer question 12.

POPULATION

⊛ Capital Cities   ⊗ 10,000 to 25,000   ⊡ 100,000 to 500,000
○ Under 5,000   ⊙ 25,000 to 50,000
⊙ 5,000 to 10,000   ⦿ 50,000 to 100,000   ▣ 500,000 and over

_____ 12. According to the information above, a city having a population of 15,000 should be indicated on a map with which symbol?

      a. ○   **b.** ⊗   **c.** ⦿   **d.** ⊡

M O D E R N   E A R T H   S C I E N C E

Chapter 3

Choose the one best response. Write the letter of that choice in the space provided.

_____ 13. On a map with a scale of 1:50,000, a measured distance of 1 cm on the map is equal to what distance on the earth's surface?

     a. 50,000 m   **b.** 5,000 m   **c.** 500 m   **d.** 50 m

_____ 14. On a topographic map, contour lines that bend to form a V-shape indicate a:

     a. hilltop.   **b.** road.   **c.** cliff.   **d.** valley.

_____ 15. A city's location on the earth with respect to the prime meridian is given by the city's:

     a. elevation.   **b.** declination.   **c.** latitude.   **d.** longitude.

_____ 16. The difference in elevation between the highest and lowest parts of an area being mapped is called the:

     a. altitude.   **b.** relief.
     c. topography.   **d.** contour interval.

_____ 17. Which of the following are shown as straight lines on a Mercator projection?

     a. compass directions   **b.** coastlines
     c. contour lines   **d.** great circles

_____ 18. A southern city in the United States has a latitude of 33°51'53'' N. The symbol **53''** represents:

     a. 53 seconds.   **b.** 53 degrees.
     c. 53 minutes.   **d.** 53 miles.

Use the diagram below to answer questions 19 and 20.

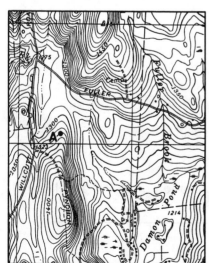

_____ 19. What is the elevation of the point labeled **A**?

     a. 1,310 feet
     b. 1,320 feet
     c. 1,330 feet
     d. 1,340 feet

_____ 20. What is the approximate elevation of the point where Fuller Brook crosses Fuller Road?

     a. 1,275 feet
     b. 1,295 feet
     c. 1,305 feet
     d. 1,325 feet

M O D E R N   E A R T H   S C I E N C E

Chapter 3

Complete each statement by writing the correct term or phrase in the space provided.

21. The line of longitude that passes through Greenwich, England, is called the

    _____ .

22. The angular distance north or south of the equator is called _____ .

23. The angle between the geographic North Pole and the direction in which a compass needle

    points is called magnetic _____ .

24. A circle formed by two meridians directly across from each other on opposite sides of the

    globe is called a _____ .

25. On a topographic map, the difference in elevation between one contour line and the next is

    the _____ .

26. Maps made by combining several conic projections are called _____ .

27. On maps, symbols for features such as cities and rivers are explained in the map's

    _____ .

28. The only line of latitude that is a great circle is the _____ .

29. On topographic map, contour lines that are printed bolder and that have their elevations

    labeled are called _____ .

30. On a topographic map, depressions are marked with loops having short, straight lines called

    _____ .

Read each question or statement and answer it in the space provided.

31. Why is a degree of longitude at the equator not equal to the same distance as a degree of
    longitude of 45° N latitude?

    _____

    _____

    _____

    _____

    _____

M O D E R N E A R T H S C I E N C E

Chapter 3

Read each question or statement and answer it in the space provided.

32. Explain how a topographic map could be useful in planning a hiking route.

_____

_____

_____

_____

_____

33. How can a magnetic compass be used to find geographic north?

_____

_____

_____

_____

_____

34. Why are gnomonic projections useful to navigators in plotting routes used in air travel?

_____

_____

_____

_____

_____

35. Interpret the location of a city that is located at 20° S, 50° W.

_____

_____

_____

_____

_____

MODERN EARTH SCIENCE

Unit 1
Studying the Earth

Read each statement below. If the statement is true, write _T_ in the space provided. If the statement is false, write _F_ in the space provided.

_____ **1.** When a chemical element is heated enough, it produces a bright-line spectrum.

_____ **2.** Rock layers containing deformed quartz particles provide important evidence for the Big Bang theory.

_____ **3.** On a Mercator projection, only the areas near 0° latitude are shown with relative accuracy.

_____ **4.** The solid rock of the asthenosphere has the ability to flow.

_____ **5.** P waves travel more slowly than S waves.

_____ **6.** Scientific research always follows the sequential steps called scientific methods.

_____ **7.** A meridian is a semicircle that runs from pole to pole.

_____ **8.** The distance covered by a degree of latitude depends upon where the degree is measured.

_____ **9.** The weight of an object on earth depends on its mass and its distance from the earth's center.

_____ **10.** During the summer solstice, the sun's vertical rays strike the earth along the Tropic of Cancer.

_____ **11.** Topographic maps of plains and other relatively flat areas have large contour intervals.

_____ **12.** Eratosthenes used shadows and simple geometry to calculate the circumference of the earth.

_____ **13.** An astronomer studies the universe beyond the earth.

_____ **14.** A satellite will stay in orbit only if there is an exact balance between its mass and its weight.

_____ **15.** The Big Bang theory proposes that the universe began in an extremely small volume.

M O D E R N   E A R T H   S C I E N C E

Unit 1

Choose the one best response. Write the letter of that choice in the space provided.

_____ **16.** The shortest distance between any two points on the globe can be found by drawing a:

        a. parallel.   **b.** great circle.
        c. prime meridian.   **d.** conic projection.

_____ **17.** The red shift observed in a distant galaxy's spectra is evidence for the:

        a. meteorite-impact hypothesis.
        b. presence of background radiation.
        c. destruction of the ozone.
        d. expansion of the universe.

_____ **18.** The ecosystem that encompasses all other ecosystems is called the:

        a. geosphere.   **b.** atmosphere.   **c.** biosphere.   **d.** hydrosphere.

_____ **19.** Which of the following best describes the magnetosphere?

        a. the region of liquid iron in the earth's outer core
        b. the region of space affected by the earth's magnetic field
        c. the region of solid iron in the earth's inner core
        d. the part of the atmosphere affected by the moon's magnetic field

_____ **20.** On a map having a scale of 1:25,000, a map distance of 2 cm represents:

        a. 50 m.   **b.** 500 m.   **c.** 5 km.   **d.** 50 km.

_____ **21.** The relationship between plants and animals in a lake would most likely be studied by:

        a. a meteorologist.   **b.** an ecologist.
        c. an oceanographer.   **d.** a geologist.

_____ **22.** The sun is highest over the center of each standard time zone at:

        a. 10:30 A.M.   **b.** 12:00 noon.   **c.** 1:30 P.M.   **d.** 3:00 P.M.

_____ **23.** Satellites in geosynchronous orbits are most often used for:

        a. communications.   **b.** surveying the earth's surface.
        c. weather information.   **d.** surveying the ocean floor.

M O D E R N   E A R T H   S C I E N C E

Unit 1

Choose the one best response. Write the letter of that choice in the space provided.

_____ **24.** The outermost layer of the earth is made up of:

    a. shadow zones and the Moho.
    b. the core and the magnetosphere.
    c. oceanic crust and continental crust.
    d. the asthenosphere.

_____ **25.** When the earth is at the farthest point in its orbit from the sun, it is said to be at:

    a. the elliptical.   **b.** perihelion.
    c. the equinox.   **d.** aphelion.

_____ **26.** Measurement instruments are most commonly used to:

    a. form a hypothesis.   **b.** state the problem.
    c. gather information.   **d.** develop a theory.

_____ **27.** On topographic maps, closely spaced contour lines indicate:

    a. shallow rivers.   **b.** gentle slopes.
    c. water.   **d.** steep slopes.

_____ **28.** On a map, a list of symbols and their meanings is called:

    a. a legend.   **b.** a scale.
    c. an index contour.   **d.** a map projection.

Complete each statement by writing the correct term or phrase in the space provided.

29. Materials that can be broken down by microorganisms into harmless substances are

_____ .

30. In scientific methods, using the senses in order to gather information is called making an

_____ .

31. The angular distance between the equator and the North Pole is _____ .

32. The general term for a vibration that travels through the earth is

_____ .

33. The time system that provides an additional hour of light during the evening is called

_____ .

M O D E R N   E A R T H   S C I E N C E

Unit 1

Complete each statement by writing the correct term or phrase in the space provided.

34. A rule that consistently describes a natural phenomenon is known as a scientific

_____ .

35. The point halfway between the highest and lowest tide levels is called

_____ .

36. A factor that can be changed in an experiment is called a _____ .

37. The imaginary line that circles the earth at 0° latitude is called the

_____ .

38. When a light source is moving toward an observer, the wavelengths of light appear

_____ .

39. The day in March when the hours of daylight and darkness are equal is called the

_____ .

40. The imaginary line running from north to south through the Pacific Ocean that indicates

where the date changes from one day to the next is called the _____ .

41. The type of map projection that shows compass directions as straight lines is a

_____ .

42. Longitude indicates the angular distance east and west of the _____ .

43. The slow, wobbling motion of the earth's axis as it turns in space is called

_____ .

44. Locations on the earth's surface where neither P waves nor S waves are detected or where

only P waves are detected are called _____ .

45. A map on which 1 cm equals 100 m has a fractional scale of _____ .

M O D E R N   E A R T H   S C I E N C E

Unit 1

Read each question or statement and answer it in the space provided.

46. Describe the distortion produced by gnomonic and conic projections.

_____

_____

_____

_____

_____

_____

_____

47. Describe the relative positions of the earth and the sun on the winter solstice.

_____

_____

_____

_____

_____

_____

_____

48. Explain the term magnetic declination.

_____

_____

_____

_____

_____

_____

_____

_____

MODERN EARTH SCIENCE

Unit 1

Read each question or statement and answer it in the space provided.

49. Describe the three major zones of the earth.

_____

_____

_____

_____

_____

_____

_____

50. What does the meteorite-impact hypothesis state?

_____

_____

_____

_____

_____

_____

M O D E R N   E A R T H   S C I E N C E

Chapter 4
Plate Tectonics

**Read each statement below. If the statement is true, write _T_ in the space provided.
If the statement is false, write _F_ in the space provided.**

_____ 1. Alfred Wegener hypothesized that continents are fixed and unable to move.

_____ 2. The similar ages and types of rocks from western Africa and eastern Brazil support the theory of continental drift.

_____ 3. The earth's oldest rocks are found along the Mid-Atlantic Ridge.

_____ 4. The theory of plate tectonics states that the earth's crust is made up of a single massive plate.

_____ 5. The Mid-Atlantic Ridge is an example of a divergent plate boundary.

_____ 6. A plate boundary at which one plate slides under another is called a transform fault boundary.

_____ 7. Island arcs form along convergent plate boundaries.

_____ 8. Many scientists believe that lithospheric plate movement is caused by temperature differences in the mantle.

_____ 9. A rift valley is an example of a suspect terrane.

_____ 10. Fossil records provide a clue to earlier positions of lithospheric plates.

Choose the one best response. Write the letter of that choice in the space provided.

_____ 11. The type of collision that occurs when two lithospheric plates converge is determined primarily by plate:

    a. density.   **b.** temperature.   **c.** size.   **d.** magnetism.

_____ 12. Paleomagnetism of the ocean floor is used to identify the:

    a. number of plate boundaries.
    b. approximate time the rock cooled and hardened.
    c. relative strength of the earth's magnetic field.
    d. kind of rock that forms the earth's crust.

_____ 13. Seafloor spreading occurs at:

    a. subduction zones.   **b.** terrane boundaries.
    c. transform faults.   **d.** divergent boundaries.

_____ 14. The region along a plate boundary where one plate is moved beneath another plate is called a:

    a. rift valley.   **b.** subduction zone.
    c. paleomagnetic band.   **d.** transform fault.

M O D E R N   E A R T H   S C I E N C E

Chapter 4

Choose the one best response. Write the letter of that choice in the space provided.

_____ **15.** As rock moves away from a mid-ocean ridge, it is replaced by:

    a. continental crust.   **b.** molten rock.
    c. suspect terranes.   **d.** older rock.

_____ **16.** The theory of suspect terranes provides an explanation of how:

    a. continents form.   **b.** magma spreads.
    c. convection currents occur.   **d.** faulting occurs.

_____ **17.** The earth's layer of solid rock that flows under pressure is called the:

    a. crust.   **b.** asthenosphere.
    c. lithosphere.   **d.** hydrosphere.

_____ **18.** Scientific evidence indicating that convection currents move the plates has been found by studying:

    a. transform faults.   **b.** mountain chains.
    c. coral atolls.   **d.** rift valleys.

Use the diagram below to answer questions 19 and 20.

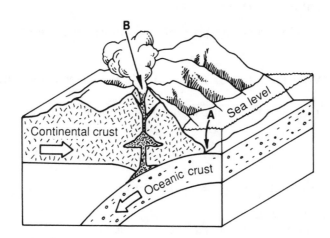

_____ **19.** Which of the following has formed at **A**?

    a. island arc   **b.** ocean trench
    c. mid-ocean ridge   **d.** magnetic band sequence

_____ **20.** What is occurring at point **B**?

    a. seafloor spreading   **b.** volcanic eruption
    c. magnetic reversal   **d.** trench development

M O D E R N   E A R T H   S C I E N C E

Chapter 4

Read each question and answer it in the space provided.

21. According to the theory of continental drift, the eastern coast of South America was once joined to what continent? _____

22. Which theory states that new land is added to continental margins at subduction zones? _____

23. Iceland is growing due to its location on what geologic feature? _____

24. Which property of iron-bearing rocks provided evidence for the hypothesis of seafloor spreading? _____

25. What two types of crust make up the surface of the earth? _____

26. What is the term for a small piece of land that is scraped off descending ocean floor at a subduction zone? _____

27. In which layer of the earth do convection currents occur? _____

28. What do scientists think is the cause of the movement of the earth's plates? _____

29. What is the name for the huge landmass that Wegener believed covered much of the earth 200 million years ago? _____

30. What type of plate boundary causes extensive volcanic activity in Iceland? _____

Read each question or statement and answer it in the space provided.

31. What hypothesis is Alfred Wegener known for, and what evidence supported his hypothesis?

_____

_____

_____

_____

_____

M O D E R N   E A R T H   S C I E N C E

Chapter 4

Read each question or statement and answer it in the space provided.

32. Describe the theory of seafloor spreading.

_____

_____

_____

_____

_____

33. What are convection currents, and how do they affect the earth's asthenosphere?

_____

_____

_____

_____

_____

34. What are the three identifying characteristics of a terrane?

_____

_____

_____

_____

_____

35. On the west coast of South America, an oceanic plate is colliding with a continental plate. According to the theory of plate tectonics, what type of geologic activity is occurring at this plate boundary? Describe the activity.

_____

_____

_____

_____

_____

M O D E R N E A R T H S C I E N C E

Chapter 5
Deformation of the Crust

Read each statement below. If the statement is true, write *T* in the space provided. If the statement is false, write *F* in the space provided.

_____ 1. Mountains that are formed when molten rock erupts violently onto the earth's surface are called dome mountains.

_____ 2. Folded mountains are commonly found where continents have collided.

_____ 3. The major cause of crustal deformation is the movement of the earth's lithospheric plates.

_____ 4. Isostatic adjustments can occur as a result of the erosion of mountain ranges.

_____ 5. Compression is the force that pulls rocks apart.

_____ 6. Plate collisions occur along converging plate boundaries.

_____ 7. Strain causes the folding and faulting of crustal rocks.

_____ 8. Anticlines generally form ridges in response to folding.

_____ 9. When land is covered by a glacier, the crust beneath it will sink lower into the mantle.

_____ 10. When motion occurs along a break in rock, the break is called a fracture.

Choose the one best response. Write the letter of that choice in the space provided.

_____ 11. A very thick accumulation of materials deposited near shore on the ocean floor causes the ocean floor to:

      a. rise.   **b.** fracture.   **c.** erupt.   **d.** sink.

_____ 12. As a volcanic mountain range is built, isostatic adjustment will cause the crust beneath it to:

      a. break.   **b.** sink.   **c.** rise.   **d.** fold.

_____ 13. The oceanic crust of the Mediterranean Sea floor is:

      a. colliding with the lithospheric plate to the east.
      b. folding due to tensional stresses.
      c. subducting beneath continental crust to the north.
      d. fracturing along strike-slip faults.

_____ 14. The structures that form where parts of the earth's crust have been broken by faults are called:

      a. grabens.   **b.** plateaus.   **c.** synclines.   **d.** monoclines.

HRW material copyrighted under notice appearing earlier in this work.

37

M O D E R N E A R T H S C I E N C E

Chapter 5

Choose the one best response. Write the letter of that choice in the space provided.

_____ 15. Folding of rocks is most likely to happen when rocks undergo:

      a. tension.   **b.** shearing.   **c.** compression.   **d.** cooling.

_____ 16. A collision between a continental plate and an oceanic plate is most likely to produce:

      a. hot spots.   **b.** mountain ranges.
      c. dome mountains.   **d.** volcanic islands.

_____ 17. Most plateaus are found:

      a. near mountain ranges.   **b.** bordering ocean basins.
      c. beneath grabens.   **d.** beside diverging boundaries.

Use the diagram below to answer questions 18-20.

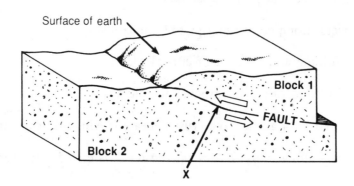

_____ 18. What type of fault is pictured in the diagram?

      a. normal   **b.** abnormal   **c.** thrust   **d.** strike-slip

_____ 19. The rocks in Block **1** form:

      a. an anticline.   **b.** a footwall.
      c. a syncline.   **d.** a hanging wall.

_____ 20. Line **X** represents a:

      a. fault plane.   **b.** collision line.
      c. subduction zone.   **d.** fold peak.

Read each question and answer it in the space provided.

21. What are the names of the earth's two mountain belts? _____

22. When rocks are exposed to stress at very low temperatures
and pressures, will they most likely undergo faulting or
folding? _____

M O D E R N   E A R T H   S C I E N C E

Chapter 5

Read each question and answer it in the space provided.

23. What type of mountain forms when molten rock rises through the crust and pushes up the rock layers above it? _____

24. What is the term for a group of adjacent mountain ranges? _____

25. What is the general term for the balancing of the up-and-down forces between the crust and the mantle? _____

26. What type of stress causes rocks to be squeezed together? _____

27. What is the general term for changes in the shape of the earth's crust? _____

28. In which type of fault do the rocks on either side of the fault plane move horizontally? _____

29. What type of mountain forms at mid-ocean ridges? _____

30. What type of mountains are the Rocky Mountains in Wyoming? _____

Read each question or statement and answer it in the space provided.

31. Explain how rivers can affect isostatic adjustment.

_____

_____

_____

_____

_____

32. Describe a reverse fault.

_____

_____

_____

_____

_____

HRW material copyrighted under notice appearing earlier in this work.

39

M O D E R N   E A R T H   S C I E N C E

Chapter 5

Read each question or statement and answer it in the space provided.

33. What kind of mountain is Mount St. Helens? Explain how this type of mountain forms.

_____

_____

_____

_____

_____

34. Describe a collision between two plates of oceanic crust.

_____

_____

_____

_____

_____

35. The San Andreas fault in California is located at a fault boundary where the Pacific plate is moving toward the northwest past the North American plate. What type of fault is the San Andreas fault? What type of stress is most likely acting on the rocks along this fault?

_____

_____

_____

_____

M O D E R N   E A R T H   S C I E N C E

Chapter 6

Earthquakes

Read each statement below. If the statement is true, write *T* in the space provided. If the statement is false, write *F* in the space provided.

_____ 1. A fault zone is a group of interconnected faults.

_____ 2. The earthquakes that cause the most damage usually have a shallow focus.

_____ 3. The New Madrid, Missouri, earthquake of 1812 resulted from movement along the San Andreas fault.

_____ 4. Scientists think that seismic gaps are likely locations of future earthquakes.

_____ 5. The magnitude of an earthquake is an approximate measure of how much energy the earthquake releases.

_____ 6. L waves are the fastest-moving seismic waves.

_____ 7. Scientists locate the epicenter of an earthquake by determining its magnitude.

_____ 8. Tsunamis are seismic waves that travel through the earth's core.

_____ 9. Most buildings are not designed to withstand the swaying motion caused by earthquakes.

_____ 10. Changes in the magnetic and electrical properties of rocks along a fault may be warnings that an earthquake will occur in the near future.

Choose the one best response. Write the letter of that choice in the space provided.

_____ 11. How far below the earth's surface do intermediate-focus earthquakes occur?

    a. 10 to 30 kilometers   **b.** 30 to 70 kilometers
    c. 70 to 300 kilometers   **d.** 300 to 650 kilometers

_____ 12. L waves are especially destructive when traveling through:

    a. water.   **b.** solid rock.   **c.** a fault zone.   **d.** loose earth.

_____ 13. The San Andreas fault zone has formed where the edge of the Pacific plate is slipping:

    a. under the North American plate.
    b. over the North American plate.
    c. south along the North American plate.
    d. north along the North American plate.

M O D E R N   E A R T H   S C I E N C E

Chapter 6

Choose the one best response. Write the letter of that choice in the space provided.

_____ **14.** Information from how many seismographs is generally needed to locate the epicenter of an earthquake?

    a. 1   **b.** 2   **c.** 3   **d.** 4

_____ **15.** P waves travel how many times faster than S waves?

    a. 0.17 times   **b.** 1.7 times   **c.** 17.0 times   **d.** 31.7 times

_____ **16.** Tests done in Colorado showed that earthquakes were less severe after faults were injected with:

    a. water.   **b.** natural gas.   **c.** rock.   **d.** loose earth.

_____ **17.** The Mercalli scale is used to express the:

    a. magnitude of an earthquake.
    b. distance of an earthquake from a seismograph.
    c. intensity of an earthquake.
    d. depth of an earthquake's focus.

_____ **18.** During the Alaskan earthquake of 1964, most deaths were caused by:

    a. the collapse of buildings.   **b.** a tsunami.
    c. vibration of the ground.   **d.** cracks formed in the earth.

Use the diagram below to answer question 19.

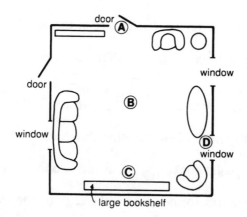

_____ **19.** During an earthquake, which position would provide the greatest safety from falling and flying debris?

    a. position **A**   **b.** position **B**   **c.** position **C**   **d.** position **D**

M O D E R N   E A R T H   S C I E N C E

Chapter 6

Use the diagram below to answer question 20.

_____ **20.** During an earthquake, which location would be safest for somebody in an automobile?

    a. location **A**   **b.** location **B**   **c.** location **C**   **d.** location **D**

Complete each statement by writing the correct term or phrase in the space provided.

21. Earthquakes take place along faults in the earth's _____ .

22. Seismic waves that can penetrate the liquid part of the earth's core are called

_____ .

23. The last waves to be recorded by a seismograph are the _____ .

24. A major earthquake must have a magnitude of at least _____ .

25. Zones of immobile rock along faults are called _____ .

26. When friction keeps a fault in an immobile state, the fault is said to be

_____ .

27. Series of smaller tremors that follow major earthquakes are called

_____ .

28. The western coast of North America is part of a major earthquake zone called the

_____ .

29. Seismographs are designed to record north-south horizontal motion, east-west horizontal

motion, and _____ .

30. The Richter scale is used to express an earthquake's _____ .

M O D E R N   E A R T H   S C I E N C E

Chapter 6

Read each question and answer it in the space provided.

31. Why do many earthquakes occur along the Mid-Atlantic Ridge?

_____

_____

_____

32. How do the Richter and Mercalli scales differ?

_____

_____

_____

33. How may faulting cause a tsunami?

_____

_____

_____

34. What is the relationship between the focus and the epicenter of an earthquake?

_____

_____

_____

35. What can you conclude about an earthquake whose P waves are recorded by a seismograph long before the S waves are recorded? Why can you reach this conclusion?

_____

_____

_____

M O D E R N   E A R T H   S C I E N C E

Chapter 7

Volcanoes

Read each statement below. If the statement is true, write _T_ in the space provided. If the statement is false, write _F_ in the space provided.

_____ 1. Knowledge of previous eruptions of a particular volcano is generally helpful in predicting its future eruptions.

_____ 2. Fissures are commonly found at the top of shield cones.

_____ 3. Magma is able to rise upward through the earth's crust because it is less dense than the surrounding rocks.

_____ 4. Tephra is usually ejected from volcanoes that erupt felsic lava.

_____ 5. The Hawaiian Islands are examples of cinder cones.

_____ 6. The temperature and pressure in the asthenosphere generally keep the rocks there below the melting point.

_____ 7. Volcanic activity is frequent in island arcs.

_____ 8. The material that erupts from volcanoes on Io is felsic.

_____ 9. The heat produced by intense meteorite bombardment may be responsible for much of the moon's ancient volcanism.

_____ 10. Cinder cones are generally much steeper than shield cones.

Choose the one best response. Write the letter of that choice in the space provided.

_____ 11. The easiest way to distinguish between volcanic ash and volcanic dust particles is to compare their:

    a. color.   **b.** weight.   **c.** diameter.   **d.** density.

_____ 12. Magma that erupts under water often forms rounded formations called:

    a. aa lava.   **b.** pillow lava.
    c. volcanic bombs.   **d.** pahoehoe lava.

_____ 13. The broad volcanic feature formed by quiet eruptions of thin lava flows is called a:

    a. shield cone.   **b.** cinder cone.
    c. rift.   **d.** stratovolcano.

_____ 14. The catastrophic volcanic eruption that caused a series of tsunamis and a drop in global temperatures happened in:

    a. Japan.   **b.** Hawaii.   **c.** Krakatau.   **d.** Iceland.

M O D E R N   E A R T H   S C I E N C E

Chapter 7

Choose the one best response. Write the letter of that choice in the space provided.

_____ **15.** One of the features supporting the theory of volcanism on the moon is the presence of:

        a. smooth crater interiors.   **b.** continued eruption today.
        c. volcanic cones.   **d.** abundant tephra.

_____ **16.** Seismographs can be useful in predicting volcanic eruptions because they measure:

        a. changes in surface bulging.   **b.** changes in gas composition.
        c. temperature increases.   **d.** earthquake activity.

_____ **17.** Which of the following is most likely to occur in an area of the asthenosphere where surrounding rock exerts less-than-normal pressure?

        a. violent volcanic eruptions   **b.** magma formation
        c. plate subduction   **d.** caldera formation

_____ **18.** As a result of the subduction of oceanic crust under a continent, magma is most likely to erupt from:

        a. an oceanic ridge.   **b.** an oceanic trench.
        c. an island arc.   **d.** a volcanic cone.

Use the diagram below to answer questions 19 and 20.

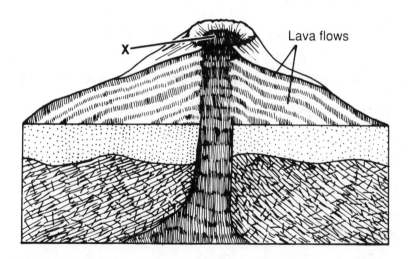

Lava flows

X

_____ **19.** What type of volcanic formation is represented by this diagram?

        a. stratovolcano   **b.** shield cone   **c.** caldera   **d.** cinder cone

_____ **20.** The feature labeled **X** is a:

        a. volcanic bomb.   **b.** hot spot.
        c. volcanic bulge.   **d.** crater.

M O D E R N   E A R T H   S C I E N C E

Chapter 7

Complete each statement by writing the correct term or phrase in the space provided.

21. The largest tephra, formed from solid rock, is known as _____ .

22. Areas of volcanism within lithospheric plates are known as _____ .

23. The composition of felsic lava differs from that of mafic lava because felsic lava contains

more _____ .

24. The volcano Olympus Mons is an unusual example of the cone type called a

_____ .

25. Lava with a wrinkled surface that forms when mafic lava hardens is known as

_____ .

26. The thin lava that generally erupts from oceanic volcanoes is called

_____ .

27. A stratovolcano is also called a _____ .

28. The funnel-shaped pit at the top of a volcanic cone that is formed by the explosion of

material is called a _____ .

29. Volcanic explosions that destroy the upper part of the cone often leave a depression called a

_____ .

30. When solid rock in the earth's mantle melts, it forms a liquid rock known as

_____ .

Read each question or statement and answer it in the space provided.

31. How are mountains formed when a plate with an oceanic crust meets one with a continental
crust?

_____

_____

_____

_____

_____

M O D E R N   E A R T H   S C I E N C E

Chapter 7

Read each question or statement and answer it in the space provided.

32. What evidence is there that volcanoes on Mars may still be active?

_____

_____

_____

_____

_____

33. How do scientists explain the volcanic activity on Jupiter's moon Io?

_____

_____

_____

_____

_____

34. Explain how new oceanic floor is formed at mid-ocean ridges.

_____

_____

_____

_____

_____

35. Explain how new islands in the Hawaiian chain may form in the future as a result of hot-spot volcanism.

_____

_____

_____

_____

_____

MODERN EARTH SCIENCE

Unit 2
The Dynamic Earth

Read each statement below. If the statement is true, write _T_ in the space provided. If the statement is false, write _F_ in the space provided.

_____ **1.** A change in the weight of part of the crust can result in crustal deformation.

_____ **2.** Collisions between oceanic and continental crust often result in the formation of volcanic mountains.

_____ **3.** A caldera is the funnel-shaped pit at the top of a volcanic vent.

_____ **4.** Stress is a force that applies pressure to the rocks in the earth's crust.

_____ **5.** Mid-ocean ridges are huge folded mountain chains.

_____ **6.** Transform faults occur at divergent plate boundaries.

_____ **7.** The theory of continental drift states that present-day continents were once a single landmass.

_____ **8.** Plate tectonics is a theory that describes the earth's surface as being in motion, with broken plates of the crust drifting slowly over the material of the asthenosphere.

_____ **9.** Bands of alternately older and younger rocks occur on each side of rifts in mid-ocean ridges.

_____ **10.** Evidence of volcanism exists in several areas of our solar system.

_____ **11.** Most earthquakes occur near plate boundaries.

_____ **12.** Scientists must first locate the epicenter of an earthquake before they can determine its distance from a seismograph.

_____ **13.** Earthquakes occur frequently near island arcs.

_____ **14.** P waves move faster through liquids than through other materials.

HRW material copyrighted under notice appearing earlier in this work.

49

M O D E R N   E A R T H   S C I E N C E

Unit 2

Choose the one best response. Write the letter of that choice in the space provided.

_____ **15.** The balance between the forces pushing the crust down and the forces pushing it up is called:

     a. strain.   **b.** isostasy.   **c.** convergence.   **d.** tectonics.

_____ **16.** Which of the following statements best describes how tension affects crustal rocks?

     a. It pulls the rocks apart.
     b. It pushes the rocks in two different directions.
     c. It squeezes the rocks together and upward.
     d. It fractures the rocks at convergent boundaries.

_____ **17.** A group of adjacent mountains with the same general shape or structure is called a mountain:

     a. system.   **b.** dome.   **c.** range.   **d.** belt.

_____ **18.** A thrust fault is a special type of:

     a. hanging wall.   **b.** footwall.
     c. reverse fault.   **d.** normal fault.

_____ **19.** What type of mountain forms when molten rock rises through the crust and pushes up the overlying rock?

     a. folded   **b.** fault-block   **c.** dome   **d.** volcanic

_____ **20.** The magnitude of an earthquake is a direct measure of:

     a. how much energy it releases.
     b. how much damage it causes.
     c. how many tsunamis it creates.
     d. how many aftershocks it causes.

_____ **21.** Estimates of the earth's inner temperatures and pressures are primarily based on the:

     a. drift rate of lithospheric plates.
     b. composition of felsic magmas.
     c. location of volcanic islands.
     d. behavior of seismic waves.

_____ **22.** Which type of seismic wave travels the fastest?

     a. L wave   **b.** P wave   **c.** S wave   **d.** surface wave

_____ **23.** Which of the following generally causes the most damage?

     a. shallow-focus earthquakes   **b.** deep-focus earthquakes
     c. intermediate-focus earthquakes   **d.** aftershocks

M O D E R N   E A R T H   S C I E N C E

Unit 2

Choose the one best response. Write the letter of that choice in the space provided.

_____ **24.** What usually happens to magma after it forms in the earth's mantle?

     a. It spreads out sideways.
     b. It sinks deeper down.
     c. It stays where it forms.
     d. It rises through cracks.

_____ **25.** The Pacific Ring of Fire is an earthquake zone that forms a ring around:

     a. the Atlantic Ocean.
     b. South America.
     c. the Pacific Ocean.
     d. North America.

_____ **26.** What geologic discovery in the 1960's provided the scientific evidence to verify the occurrence of continental drift?

     a. fossils of the same plants and animals on continental coasts
     b. eroded land bridges that once connected continents
     c. coal deposits in the United States, Europe, and Siberia
     d. reversed magnetic polarities in rocks on land and ocean floors

_____ **27.** Where do convection currents in the earth occur?

     a. in the asthenosphere
     b. in oceanic crust
     c. in the lithosphere
     d. in continental crust

_____ **28.** Two plates moving away from each other form a:

     a. convergent boundary.
     b. subduction zone.
     c. divergent boundary.
     d. transform fault boundary.

_____ **29.** A seismic gap is a region in which:

     a. there are no seismographs.
     b. tsunamis never occur.
     c. a fault is locked.
     d. only deep-focus earthquakes occur.

_____ **30.** What happens when a plate with oceanic crust meets a plate with continental crust?

     a. The continental plate is subducted.
     b. New crust forms over a hot spot.
     c. The oceanic plate is subducted.
     d. New crust forms under a hot spot.

M O D E R N   E A R T H   S C I E N C E

Unit 2

Complete each statement by writing the correct term or phrase in the space provided.

31. A group of interconnected faults at a plate boundary is called a

_____ .

32. Scientists plot three intersecting circles to locate an earthquake's _____ .

33. Island arcs are found where the collision of two oceanic plates has resulted in the formation

of a deep _____ .

34. The movement of the ocean floor away from a mid-ocean ridge is called

_____ .

35. Pieces of land bordered by faults that have geologic features different from those of

neighboring pieces of land are called _____ .

36. When lava flows out of fissures on the ocean floor, it produces a formation called

_____ .

37. The largest pieces of volcanic material formed from solid rock blasted from a volcano are

called _____ .

38. The movement of magma toward the surface of the earth is called

_____ .

39. Narrow valleys that form in the centers of mid-ocean ridges as plates separate are called

_____ .

40. The region along a plate boundary where one plate is forced under another plate is called a

_____ .

41. Many scientists attribute the movement of lithospheric plates to the process of heat transfer

called _____ .

42. The layer of plastic rock below the lithosphere is called the _____ .

M O D E R N   E A R T H   S C I E N C E

Unit 2

Complete each statement by writing the correct term or phrase in the space provided.

43. A fold that is characterized by a gently dipping bend in horizontally layered rocks is called a

_____ .

44. A very thick accumulation of sand, gravel, and rock material deposited by a river on the

floor of the ocean causes it to _____ .

Read each question or statement and answer it in the space provided.

45. Describe the parts of a normal fault.

_____

_____

_____

_____

_____

46. Why are small earthquakes important warning signals of volcanic eruptions?

_____

_____

_____

_____

_____

47. Describe the relationship between the lithosphere and the asthenosphere in terms of plate tectonics.

_____

_____

_____

_____

_____

M O D E R N   E A R T H   S C I E N C E

Unit 2

Read each question or statement and answer it in the space provided.

48. How does an earthquake occur?

_____

_____

_____

_____

_____

49. Why are eruptions of oceanic volcanoes usually quieter than eruptions of continental volcanoes?

_____

_____

_____

_____

_____

50. Describe the two events that generally cause tsunamis.

_____

_____

_____

_____

_____

Chapter 8
Earth Chemistry

Read each statement below. If the statement is true, write *T* in the space provided. If the statement is false, write *F* in the space provided.

_____ 1. An element may be made of more than one basic type of atom.

_____ 2. The mass number of an atom is the sum of the number of protons and electrons in that atom.

_____ 3. Protons and neutrons form the nucleus of an atom.

_____ 4. In general, the particles that make up a gas move faster than those that make up a solid.

_____ 5. Water is an example of a diatomic molecule.

_____ 6. Atoms in which the outermost energy level is filled form compounds easily with other elements.

_____ 7. Metals are generally good electrical conductors.

_____ 8. A covalent bond is formed by the transfer of electrons.

_____ 9. Electrons are shared in covalent compounds.

_____ 10. Atoms of any one element are different from atoms of all other elements.

Choose the one best response. Write the letter of that choice in the space provided.

_____ 11. Approximately how many elements occur naturally in the earth?

      a. 12   **b.** 75   **c.** 90   **d.** 120

_____ 12. An electron is a type of:

      a. isotope.   **b.** element.
      c. subatomic particle.   **d.** chemical bond.

_____ 13. An atom that contains 3 protons, 4 neutrons, and 3 electrons has an atomic number of:

      a. 3.   **b.** 4.   **c.** 7.   **d.** 10.

M O D E R N   E A R T H   S C I E N C E

Chapter 8

Choose the one best response. Write the letter of that choice in the space provided.

_____ **14.** Atoms of the same element that have different numbers of neutrons are called:

      a. alloys.   **b.** compounds.   **c.** isotopes.   **d.** ions.

_____ **15.** Which of the following has no positive or negative charge?

      a. electron   **b.** proton   **c.** ion   **d.** neutron

_____ **16.** Which of the following is shared by atoms in a covalent bond?

      a. electrons   **b.** protons   **c.** ions   **d.** neutrons

_____ **17.** How many atoms of sulfur (S) are in one molecule of H_2SO_4?

      a. 1   **b.** 2   **c.** 4   **d.** 6

_____ **18.** Seawater, smog, and brass are all examples of:

      a. alloys.   **b.** mixtures.
      c. ionic compounds.   **d.** covalent compounds.

Use the diagram below to answer questions 19 and 20.

Noble Gases

			4.00 ^{2} **He** Helium 2
14.01 $^2_5$ **N** Nitrogen 7	16.00 $^2_6$ **O** Oxygen 8	19.00 $^2_7$ **F** Fluorine 9	20.18 $^2_8$ **Ne** Neon 10
30.97 $^{2}_{8,5}$ **P** Phosphorus 15	32.07 $^2_{8,6}$ **S** Sulfur 16	35.45 $^2_{8,7}$ **Cl** Chlorine 17	39.95 $^2_{8,8}$ **Ar** Argon 18

_____ **19.** What is the mass number of the element fluorine (F)?

      a. 9   **b.** 19   **c.** 27   **d.** 28

_____ **20.** How many neutrons are in one atom of the element fluorine (F)?

      a. 9   **b.** 10   **c.** 19   **d.** 27

M O D E R N   E A R T H   S C I E N C E

Chapter 8

Complete each statement by writing the correct term or phrase in the space provided.

21. Substances that cannot be broken down into simpler forms by ordinary chemical means are

    called _____ .

22. The form of matter that has no definite shape or volume is a _____ .

23. The term that scientists use to describe anything that takes up space and has mass is

    _____ .

24. The most stable atoms have filled _____ .

25. A substance with a definite volume but no definite shape is a _____ .

26. A chlorine atom that carries a charge is called a chlorine _____ .

27. In an atom, the region of space in which electrons move is called the

    _____ .

28. The most basic subatomic particles known to scientists are leptons and

    _____ .

29. Solid solutions of two or more metals are called _____ .

30. The chart used to classify the elements is called the _____ .

Read each question or statement and answer it in the space provided.

31. Describe what happens to a solid when it is melted by heat.

    _____

    _____

    _____

    _____

    _____

MODERN EARTH SCIENCE

Chapter 8

Read each question or statement and answer it in the space provided.

32. Explain how isotopes of the same element differ from each other.

_____

_____

_____

_____

_____

33. Explain the meanings of the terms mixture and solution.

_____

_____

_____

_____

_____

34. Why is the melting point of ice a physical property rather than a chemical property?

_____

_____

_____

_____

_____

35. Using the chemical formula H_2O, explain the meaning of the letters and numbers that make up a chemical formula.

_____

_____

_____

_____

_____

M O D E R N   E A R T H   S C I E N C E

Chapter 9
Minerals of the Earth's Crust

Read each statement below. If the statement is true, write *T* **in the space provided. If the statement is false, write** *F* **in the space provided.**

_____ 1. A mineral is any solid, crystalline substance that occurs naturally.

_____ 2. All rock-forming minerals contain atoms of silicon (Si) and oxygen (O).

_____ 3. Feldspars are minerals composed of atoms of silicon (Si), oxygen (O), and a metal.

_____ 4. Feldspars and quartz make up more than 50 percent of the earth's crust.

_____ 5. Fluorite (CaF_2) is an example of a nonsilicate mineral.

_____ 6. All carbonates contain carbon and silicon atoms.

_____ 7. Nonsilicate minerals can contain elements such as carbon, gold, and iron.

_____ 8. Color is the most reliable indicator of the identity of a mineral.

_____ 9. Oil and coal are classified as silicate minerals.

_____ 10. In network silicates, each tetrahedron in the mineral is bonded to four other tetrahedra.

Choose the one best response. Write the letter of that choice in the space provided.

_____ 11. The ten most common minerals make up what percentage of the earth's crust?

      a. 90%   **b.** 50%   **c.** 20%   **d.** 10%

_____ 12. The properties of a mineral are primarily a result of its:

      a. hardness.   **b.** density.
      c. chemical composition.   **d.** cleavage pattern.

_____ 13. The way the surface of a mineral reflects light is called the mineral's:

      a. luster.   **b.** luminescence.   **c.** reflectance.   **d.** streak.

_____ 14. How many silicon atoms are in a basic silicon-oxygen tetrahedron?

      a. 1   **b.** 2   **c.** 3   **d.** 4

_____ 15. Minerals composed of silicon-oxygen tetrahedra linked by atoms of elements other than silicon and oxygen are called:

      a. sheet silicates.   **b.** chain silicates.
      c. network silicates.   **d.** ionic silicates.

M O D E R N   E A R T H   S C I E N C E

Chapter 9

Choose the one best response. Write the letter of that choice in the space provided.

_____ **16.** A mineral that splits into even sheets shows which of the following properties?

    a. low density   **b.** consistent streak
    c. good cleavage   **d.** triclinic crystal shape

_____ **17.** The ability of a mineral to glow during and after exposure to ultraviolet light is called:

    a. phosphorescence.   **b.** luminescence.
    c. fluorescence.   **d.** radioactivity.

_____ **18.** The most common magnetic mineral is:

    a. hematite.   **b.** magnetite.   **c.** halite.   **d.** uranium.

Use the diagrams below to answer questions 19 and 20.

      I   II   III   IV   V

_____ **19.** Which of the following describes the shape of crystal I?

    a. monoclinic   **b.** isometric   **c.** triclinic   **d.** hexagonal

_____ **20.** Which diagram represents an orthorhombic crystal?

    a. II   **b.** III   **c.** IV   **d.** V

Read each question and answer it in the space provided.

21. What is the formula for calculating the density of a
mineral? _____

22. What is the most common silicate mineral? _____

23. Which property is used to identify the color of a mineral
in powdered form? _____

24. In network silicates, how many of the oxygen atoms in
the silicon-oxygen tetrahedron are shared with other
tetrahedra? _____

Chapter 9

Read each question and answer it in the space provided.

25. What type of crystalline structure do pyroxenes have?

_____

26. What special property does the mineral pitchblende have?

_____

27. What is the term for the bending of light rays as they pass through a mineral?

_____

28. What term is used to describe a curved surface on a fractured mineral?

_____

29. What is the name for elements that are commonly found uncombined with other elements?

_____

30. What is the term for the common minerals that make up the earth's crust?

_____

Read each question or statement and answer it in the space provided.

31. Explain why quartz is harder than feldspar.

_____

_____

_____

_____

_____

_____

32. What are the four basic questions scientists ask about a substance to determine whether it is a mineral or a nonmineral?

_____

_____

_____

_____

_____

MODERN EARTH SCIENCE

Chapter 9

Read each question or statement and answer it in the space provided.

33. Describe the property of cleavage in minerals. Name the factor that controls the cleavage of a mineral.

_____

_____

_____

_____

_____

_____

34. Compare the jobs of a mining inspector and a mining engineer.

_____

_____

_____

_____

_____

35. How is the Mohs hardness scale used to help mineralogists determine the identity of unknown minerals?

_____

_____

_____

_____

_____

M O D E R N   E A R T H   S C I E N C E

Chapter 10
Rocks

**Read each statement below. If the statement is true, write _T_ in the space provided.
If the statement is false, write _F_ in the space provided.**

_____ 1. A porphyry is an igneous rock with a mixture of small and large grains.

_____ 2. Metamorphic rock may form from preexisting metamorphic rock.

_____ 3. On the moon's surface, maria are formed by sedimentary rock.

_____ 4. Limestone can form as a chemical sedimentary rock or as an organic sedimentary rock.

_____ 5. Conglomerates are formed from angular, gravel-sized fragments with sharp corners.

_____ 6. Cavities in sandstone that contain quartz or calcite crystals are called geodes.

_____ 7. Volcanism is the changing of one type of rock to another by heat, pressure, and chemical processes.

_____ 8. Metamorphism that occurs over an area of thousands of square kilometers is called contact metamorphism.

_____ 9. Unfoliated metamorphic rocks do not contain layers of crystals.

_____ 10. Igneous rock is produced when magma cools.

Choose the one best response. Write the letter of that choice in the space provided.

_____ 11. Gabbro is an example of a:

    a. felsic, fine-grained rock.   **b.** felsic, coarse-grained rock.
    c. mafic, fine-grained rock.   **d.** mafic, coarse-grained rock.

_____ 12. Magma that cools deep below the earth's crust forms what type of rock?

    a. clastic   **b.** intrusive   **c.** stratified   **d.** extrusive

_____ 13. A layer of sedimentary rock with coarse grains at the bottom and fine grains at the top is:

    a. foliation.   **b.** concretion.
    c. graded bedding.   **d.** cross-bedding.

_____ 14. Which of the following is classified as a metamorphic rock?

    a. basalt   **b.** diorite   **c.** limestone   **d.** schist

HRW material copyrighted under notice appearing earlier in this work.

63

Chapter 10

Choose the one best response. Write the letter of that choice in the space provided.

_____ **15.** Ripple marks in sandstone may form by the action of:

      a. wind.   **b.** magma.   **c.** heat.   **d.** intrusion.

_____ **16.** Which of the following form the core of many major mountain ranges?

      a. conglomerates   **b.** concretions
      c. batholiths   **d.** extrusions

_____ **17.** Where does most metamorphic rock form?

      a. deep below the earth's surface   **b.** within volcanoes
      c. on the earth's surface   **d.** on lake beds

_____ **18.** Which of the following is an organic sedimentary rock?

      a. basalt   **b.** coal   **c.** conglomerate   **d.** sandstone

Use the diagram below to answer questions 19 and 20.

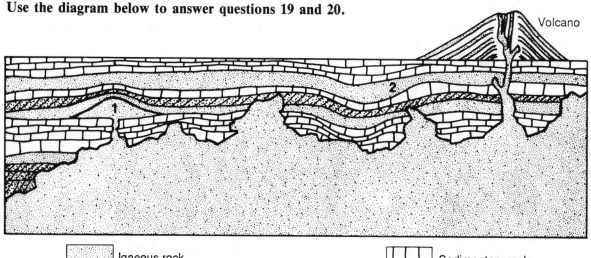

Igneous rock   Sedimentary rock

_____ **19.** The feature labeled **1** is an example of a:

      a. lava plateau.   **b.** laccolith.   **c.** geode.   **d.** batholith.

_____ **20.** The feature labeled **2** is an example of a:

      a. sill.   **b.** dike.   **c.** plug.   **d.** neck.

M O D E R N   E A R T H   S C I E N C E

Chapter 10

Complete each statement by writing the correct term or phrase in the space provided.

21. The texture of igneous rock is largely determined by the rate at which magma

_____ .

22. The process during which pressure forces sediment fragments together and squeezes out air and water from between the fragments is called _____ .

23. Remains or traces of ancient plants and animals that are preserved in rock are called

_____ .

24. Minerals precipitated from solution that build up around existing rock particles form structures called _____ .

25. Rocks are classified into three groups based on the way the rocks are

_____ .

26. Obsidian forms through the extremely rapid cooling of magma, which prevents the formation of _____ .

27. Small dome mountains such as the Black Hills of South Dakota are formed by intrusions called _____ .

28. The solidified central vent of an ancient volcano is called a volcanic

_____ .

29. Clastic sedimentary rock formed from angular fragments is called a

_____ .

30. When sediment deposition occurs in curved slopes, the inclined layers are said to be

_____ .

Read each question or statement and answer it in the space provided.

31. How may foliation occur in metamorphic rocks?

_____

_____

_____

_____

M O D E R N   E A R T H   S C I E N C E

Chapter 10

Read each question or statement and answer it in the space provided.

32. Why do most extrusive igneous rocks have small mineral crystals?

_____

_____

_____

_____

33. How is clastic sedimentary rock formed?

_____

_____

_____

_____

34. How are lava plateaus formed?

_____

_____

_____

_____

35. Explain how igneous rock can change into sedimentary rock.

_____

_____

_____

_____

M O D E R N   E A R T H   S C I E N C E

Chapter 11
Resources and Energy

Read each statement below. If the statement is true, write *T* in the space provided. If the statement is false, write *F* in the space provided.

_____ **1.** Fossil fuels are available in limitless supplies.

_____ **2.** Energy obtained directly from heat in the earth's crust is called geothermal energy.

_____ **3.** Placer deposits are formed by contact metamorphism.

_____ **4.** The process of splitting the nucleus of a large atom into two or more smaller nuclei is called nuclear fission.

_____ **5.** Anthracite is a light-colored mineral commonly used as a gemstone.

_____ **6.** Halite, sulfur, and diamond are nonmetallic minerals used as building materials.

_____ **7.** Microorganisms that lived in ancient oceans and lakes are an important source of hydrocarbons.

_____ **8.** All of the world's energy needs could easily be met by modern wind-driven generators.

_____ **9.** Sulfur dioxide is a pollutant that is commonly produced when fossil fuels are burned.

_____ **10.** Petroleum is used to produce plastics, detergents, and some medicines and insecticides.

Choose the one best response. Write the letter of that choice in the space provided.

_____ **11.** Magnetite and hematite are important sources of:

    a. mercury.   **b.** lead.   **c.** iron.   **d.** copper.

_____ **12.** Which of the following is produced by the process of carbonization?

    a. gold   **b.** coal   **c.** quartz   **d.** cinnabar

_____ **13.** Nonmetallic minerals prized for their brilliance and color are called:

    a. metals.   **b.** placers.   **c.** ores.   **d.** gemstones.

MODERN EARTH SCIENCE

Chapter 11

Choose the one best response. Write the letter of that choice in the space provided.

_____ **14.** Which of the following minerals is obtained from bauxite?

         a. gold   **b.** aluminum   **c.** platinum   **d.** sulfur

_____ **15.** Bituminous coal is produced when extreme pressure is applied to:

         a. lignite.   **b.** anthracite.   **c.** natural gas.   **d.** crude oil.

_____ **16.** In general, the most efficient conductors of heat and electricity are:

         a. metals.   **b.** petrochemicals.
         c. gemstones.   **d.** ores.

_____ **17.** Using a rooftop collector to capture the sun's energy is an example of:

         a. nuclear fusion.   **b.** nuclear fission.
         c. active solar heating.   **d.** passive solar heating.

_____ **18.** Recently, scientists have focused sunlight onto liquid saltpeter in order to make use of:

         a. nuclear energy.   **b.** geothermal energy.
         c. solar energy.   **d.** hydroelectric energy.

Use the diagram below to answer questions 19 and 20.

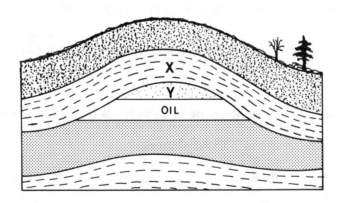

_____ **19.** Which of the following is most likely to be found in the layer labeled **Y**?

         a. gasoline   **b.** water   **c.** natural gas   **d.** peat

_____ **20.** The layer labeled **X** is most likely composed of:

         a. lignite.   **b.** sandstone.   **c.** bauxite.   **d.** shale.

M O D E R N   E A R T H   S C I E N C E

Chapter 11

Complete each statement by writing the correct term or phrase in the space provided.

21. A narrow, fingerlike band of a mineral is called a _____ .

22. Fossil fuels are made up of compounds of carbon and hydrogen

called _____ .

23. Chromium, nickel, and lead ores commonly form underground within cooling

_____ .

24. The rock immediately above a deposit of petroleum is called a _____ .

25. Fossil fuels are found in the earth in the form of crude oil, or unrefined

_____ .

26. The sun's energy that reaches the earth is produced by the process of nuclear

_____ .

27. In the future, nuclear fusion reactors may be fueled by an almost limitless supply of hydrogen

atoms from the _____ .

28. In a geothermal generating plant, energy is produced when steam turns a

_____ .

29. Chemicals derived from petroleum are called _____ .

30. Fuel rods in a nuclear reactor are made from isotopes of the

element _____ .

Read each question or statement and answer it in the space provided.

31. Describe how substitution and recycling can be used to conserve mineral resources.

_____

_____

_____

_____

_____

M O D E R N   E A R T H   S C I E N C E

Chapter 11

Read each question or statement and answer it in the space provided.

32. Explain how strip mining can pose a threat to life in nearby rivers and streams.

_____

_____

_____

_____

_____

33. Explain how movements of ocean water can be used to produce electricity.

_____

_____

_____

_____

_____

34. Explain why coal is a more efficient heat source than peat.

_____

_____

_____

_____

_____

35. In nuclear reactors, the rate of the fission reaction is controlled by raising and lowering rods made from materials that absorb neutrons. How do these rods control the nuclear fission reaction?

_____

_____

_____

_____

_____

M O D E R N   E A R T H   S C I E N C E

Unit 3
Composition of the Earth

Read each statement below. If the statement is true, write *T* in the space provided. If the statement is false, write *F* in the space provided.

_____ **1.** Electrons and neutrons are kinds of subatomic particles.

_____ **2.** A material that contains two or more substances that are not chemically combined is called a mixture.

_____ **3.** Mohs Scale is used to describe the density of minerals.

_____ **4.** All minerals occur naturally in the earth.

_____ **5.** All minerals have the same crystal shape.

_____ **6.** Large rocks have a higher density than small rocks.

_____ **7.** Fossils are commonly found in gabbro.

_____ **8.** Magma that cools quickly produces rocks that have a fine-grained texture.

_____ **9.** All metamorphic rocks come from sedimentary rocks.

_____ **10.** Dikes are produced by volcanic activity.

_____ **11.** Air and water are nonrenewable resources.

_____ **12.** Natural gas is formed from the remains of living things.

_____ **13.** Fuel rods for nuclear reactors contain two isotopes of uranium.

_____ **14.** Technology for using geothermal energy is not yet available.

Choose the one best response. Write the letter of that choice in the space provided.

_____ **15.** An atom that carries an electrical charge is called an:

    a. isotope.   **b.** ion.   **c.** element.   **d.** alloy.

_____ **16.** How many atoms of oxygen are in one molecule of H_2SO_4?

    a. 1   **b.** 2   **c.** 4   **d.** 6

_____ **17.** An atom that is chemically unreactive is most likely to have how many electrons?

    a. 1   **b.** 5   **c.** 10   **d.** 12

HRW material copyrighted under notice appearing earlier in this work.

71

M O D E R N   E A R T H   S C I E N C E

Unit 3

Choose the one best response. Write the letter of that choice in the space provided.

_____ **18.** A substance that cannot be broken down into a simpler form by ordinary chemical means is called:

      a. an element.   **b.** a compound.
      c. a mixture.   **d.** an isotope.

_____ **19.** The parent material for all rocks is:

      a. silica.   **b.** quartz.
      c. granite.   **d.** magma.

_____ **20.** The size and arrangement of crystalline grains in igneous rock is called:

      a. density.   **b.** texture.
      c. hardness.   **d.** luster.

_____ **21.** The changing of rock by pressure from colliding tectonic plates is called:

      a. igneous intrusion.   **b.** igneous extrusion.
      c. contact metamorphism.   **d.** regional metamorphism.

Use the table below to answer questions 22 and 23.

PROPERTIES OF MINERALS

Mineral	Hardness	Streak	Color	Luster
augite	5–6	greenish-gray	green to black	nonmetallic
garnet	6.5–7.5	white	dark red	nonmetallic
magnetite	5–6	black	iron black	metallic
pyrite	6–6.5	green to black	brass yellow	metallic
sphalerite	3.5	reddish-brown	brown to black	nonmetallic

_____ **22.** Which of the following minerals is white in powdered form?

      a. garnet   **b.** pyrite   **c.** augite   **d.** sphalerite

_____ **23.** Which of the following minerals can scratch pyrite?

      a. sphalerite   **b.** augite   **c.** magnetite   **d.** garnet

M O D E R N   E A R T H   S C I E N C E

Unit 3

Choose the one best response. Write the letter of that choice in the space provided.

_____ **24.** Hot mineral solutions flowing through cracks in rocks are most likely to produce:

      a. veins.   **b.** placer deposits.
      c. peat.   **d.** crude oil.

_____ **25.** The sun produces its high surface temperature through which type of energy?

      a. geothermal   **b.** nuclear fusion
      c. hydroelectric   **d.** nuclear fission

_____ **26.** Which of the following produces the largest amount of heat when it burns?

      a. peat   **b.** brown coal
      c. anthracite   **d.** soft coal

_____ **27.** A mineral in which all atoms are bonded tightly together will most likely be:

      a. metallic.   **b.** hard.
      c. dense.   **d.** magnetic.

Complete each statement by writing the correct term or phrase in the space provided.

28. A subatomic particle that has no electric charge is called a _____ .

29. An atom that has 11 neutrons and an atomic number of 10 has a mass number of

_____ .

30. The physical state of matter in which particles are packed tightly together in fixed positions is

a _____ .

31. Whether an element is a metal or a nonmetal is determined by the number and arrangement

of an atom's _____ .

32. The two main types of minerals are _____ .

33. A curved surface on a fractured mineral is called _____ .

34. A mineral sample that has a mass of 75 g and a volume of 25 cm³ has a density of

_____ .

M O D E R N   E A R T H   S C I E N C E

Unit 3

Complete each statement by writing the correct term or phrase in the space provided.

35. A natural, inorganic, crystalline solid is called a _____ .

36. Elements that have unstable arrangements of neutrons and protons are called

_____ .

37. An igneous intrusion that lies parallel to surrounding rock layers is called a

_____ .

38. Dark-colored igneous rocks that contain little silica are called _____ .

39. Texture of igneous rocks is determined primarily by the rate at which the lava

_____ .

40. A batholith-like structure that covers an area less than 100 km² is called a

_____ .

41. A sedimentary rock that contains quartz or calcite crystals in a hollow core is a

_____ .

42. Fossil fuels consist primarily of compounds of carbon and hydrogen called

_____ .

43. Deposits from which minerals can be profitably removed are called

_____ .

44. Petroleum is produced when organic matter is subjected to great heat and

_____ .

45. In nuclear fission reactors, atomic nuclei are split when they are struck by

_____ .

46. Natural gas is often found immediately above deposits of _____ .

M O D E R N   E A R T H   S C I E N C E

Unit 3

Read each question or statement and answer it in the space provided.

47. Explain what is meant by the chemical and physical properties of matter.

_____

_____

_____

_____

_____

48. Describe fluorescence and phosphorescence.

_____

_____

_____

_____

_____

49. Explain how compaction and cementation form sedimentary rock.

_____

_____

_____

_____

_____

50. How does the process of carbonization produce coal from the remains of plants?

_____

_____

_____

_____

_____

M O D E R N   E A R T H   S C I E N C E

Chapter 12
Weathering and Erosion

Read each statement below. If the statement is true, write _T_ in the space provided. If the statement is false, write _F_ in the space provided.

_____ **1.** Joints are often formed when the pressure on rock decreases.

_____ **2.** Rocks weather most rapidly in a hot, dry climate.

_____ **3.** Tropical soils generally make good farmland.

_____ **4.** Steep slopes are often too dry to support plant growth.

_____ **5.** Transported soils usually have three distinct horizons.

_____ **6.** Galvanizing is a process used to inhibit rusting.

_____ **7.** The C horizon in a residual soil consists of partially weathered bedrock.

_____ **8.** Mudflows occur most often on plains and peneplains.

_____ **9.** The most rapid and sudden type of downslope mass movement is known as creep.

_____ **10.** Oxidation often produces a reddish soil.

Choose the one best response. Write the letter of that choice in the space provided.

_____ **11.** Which of the following is a type of mechanical weathering?

        a. oxidation   **b.** hydrolysis   **c.** carbonation   **d.** ice wedging

_____ **12.** Which of the following generally weathers most rapidly?

        a. limestone   **b.** obsidian   **c.** granite   **d.** gneiss

_____ **13.** Which of the following minerals is most resistant to weathering?

        a. calcite   **b.** feldspar   **c.** sandstone   **d.** quartz

_____ **14.** Glaciers weather surface rocks mainly through the process of:

        a. oxidation.   **b.** carbonation.
        c. abrasion.   **d.** decomposition.

_____ **15.** Which of the following soil types has the smallest particle size?

        a. silt   **b.** sand   **c.** clay   **d.** gravel

_____ **16.** Which of the following is a type of accelerated erosion?

        a. leaching   **b.** terracing   **c.** gullying   **d.** wedging

M O D E R N   E A R T H   S C I E N C E

Chapter 12

Choose the one best response. Write the letter of that choice in the space provided.

_____ **17.** Humus is made up mainly of:

        a. weathered rock fragments.   **b.** organic material.

        c. unjointed bedrock.   **d.** iron particles.

_____ **18.** Which of the following rocks is most likely to weather quickly?

        a. an exposed rock on a plain   **b.** an exposed rock on a slope

        c. a buried rock on a plain   **d.** a buried rock on a slope

Use the diagram below to answer questions 19 and 20.

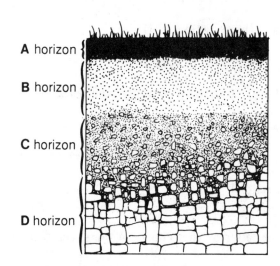

_____ **19.** Which horizon in the diagram accumulates leached minerals?

      a. A   **b. B**   **c. C**   **d. D**

_____ **20.** Which layer in the diagram represents the soil's parent rock?

      a. A   **b. B**   **c. C**   **d. D**

Read each question and answer it in the space provided.

21. Which type of mechanical weathering occurs when water seeps into cracks in a rock and freezes?   _____

22. What common mineral in limestone is most easily weathered by carbonation?   _____

23. In what type of decomposition reaction do hydrogen ions from water displace elements in a mineral?   _____

24. What is the name for a soil containing iron and aluminum minerals that do not dissolve easily in water?   _____

M O D E R N   E A R T H   S C I E N C E

Chapter 12

Read each question and answer it in the space provided.

25. What is the name of the process by which the products of weathering are transported?

_____

26. Which type of accelerated erosion occurs when high winds or the removal of cover plants causes parallel layers of topsoil to be stripped away?

_____

27. Which type of mass movement occurs when a large block of soil and rock slides downhill along a curved slope due to gravity?

_____

28. What is the correct term for a knob of rock that rises up from a peneplain?

_____

29. What type of mechanical weathering is commonly caused by wind-carried sand?

_____

30. What is the name for the landform that is created when a plateau is dissected into smaller tablelike areas?

_____

Read each question or statement and answer it in the space provided.

31. How can youthful mountains be distinguished from mature mountains?

_____

_____

_____

_____

_____

_____

32. What are pedocal soils, and under what circumstances do they form?

_____

_____

_____

_____

_____

MODERN EARTH SCIENCE

Chapter 12

Read each question or statement and answer it in the space provided.

33. How would a soil formed from a feldspar-rich rock differ from a soil formed from a quartz-rich rock?

_____

_____

_____

_____

_____

_____

34. Describe three farming methods that are used to reduce the erosion of farmland.

_____

_____

_____

_____

_____

_____

35. How is a desert soil similar to an arctic soil?

_____

_____

_____

_____

_____

_____

M O D E R N   E A R T H   S C I E N C E

Chapter 13
Water and Erosion

Read each statement below. If the statement is true, write *T* in the space provided. If the statement is false, write *F* in the space provided.

_____ **1.** The continuous movement of water between the atmosphere and the earth's surface is called the hydrologic cycle.

_____ **2.** The decline in the prosperity of Bruges, Belgium is due to the process of stream deposition.

_____ **3.** A divide is the area separating a stream's bed load from its suspended load.

_____ **4.** Stream piracy decreases the size of a river system.

_____ **5.** In a suspended stream load, sediment advances by sliding and rolling.

_____ **6.** The tip of a fan-shaped deposit at the mouth of a stream usually points upstream.

_____ **7.** Most water used by industry is returned to rivers and oceans as waste water.

_____ **8.** Feeder streams are also called tributaries.

_____ **9.** In areas of harsh winters, spring floods are common near headwaters.

_____ **10.** Dam building is the most common method of direct flood control.

Choose the one best response. Write the letter of that choice in the space provided.

_____ **11.** Which of the following creates a water gap?

    a. headward erosion   **b.** stream piracy
    c. uplifted land area   **d.** increased evaporation

_____ **12.** Artificial levees are most commonly used to prevent:

    a. evaporation.   **b.** flooding.   **c.** runoff.   **d.** gullying.

_____ **13.** Alluvial fans and deltas have similar:

    a. causes for deposition.   **b.** sizes of sediment grains.
    c. shapes of deposition.   **d.** angles of slope.

_____ **14.** One of the principal components of the sediment carried in a bed load is:

    a. mud.   **b.** silt.   **c.** sand.   **d.** gravel.

Choose the one best response. Write the letter of that choice in the space provided.

_____ **15.** A river can be rejuvenated by:

        a. increasing slope.   **b.** decreasing velocity.
        c. increasing branching.   **d.** decreasing depth.

_____ **16.** Which of the following describes the trend in the water-movement pattern going from the equator to the poles?

        a. condensation decreases   **b.** evapotranspiration decreases
        c. condensation increases   **d.** evapotranspiration increases

_____ **17.** As a river passes from a youthful to a mature stage, it:

        a. shortens.   **b.** deepens.   **c.** meanders.   **d.** narrows.

_____ **18.** The amount of precipitation on earth is equal to the amount of evapotranspiration and:

        a. runoff.   **b.** headwater.   **c.** discharge.   **d.** groundwater.

_____ **19.** Approximately what percentage of all precipitation on earth falls on the oceans?

        a. 10%   **b.** 25%   **c.** 50%   **d.** 75%

Use the diagram below to answer question 20.

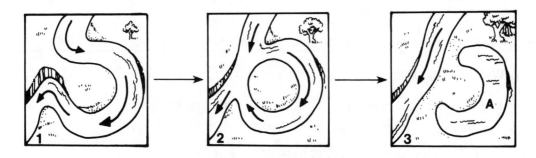

_____ **20.** In this diagram, the feature labeled **A** represents:

        a. a meander.   **b.** an oxbow lake.
        c. a pothole.   **d.** an alluvial fan.

Complete each statement by writing the correct term or phrase in the space provided.

21. The balance between precipitation on one side and evapotranspiration and runoff on the other

in a particular area is called the _____ .

22. When water vapor rises in the atmosphere, it expands and _____ .

M O D E R N   E A R T H   S C I E N C E

Chapter 13

Complete each statement by writing the correct term or phrase in the space provided.

23. The process of removing salt from ocean water is called _____ .

24. Water running across surface sediments erodes a narrow ditch called

    _____ .

25. The accumulation of flood deposits along the banks of a stream eventually produces raised

    banks called _____ .

26. The short jumps by which rocks move in a river bed are called _____ .

27. The Cumberland Gap is an example of a _____ .

28. The path that a stream follows is called its _____ .

29. The part of the valley floor that may be covered with water during a flood is called

    _____ .

30. A term for the artificial channel that carries away excess water from the main stream is

    _____ .

Read each question or statement and answer it in the space provided.

31. What are the factors that affect the local water budget?

    _____

    _____

    _____

    _____

32. Describe the processes of evaporation and transpiration.

    _____

    _____

    _____

    _____

M O D E R N   E A R T H   S C I E N C E

Chapter 13

Read each question or statement and answer it in the space provided.

33. Describe the changes in a stream's gradient and velocity as it moves from its headwaters to its mouth.

_____

_____

_____

_____

34. How does a stream's velocity affect the total load it can carry?

_____

_____

_____

_____

35. Explain how the Mississippi River can be increasing the land area of the United States.

_____

_____

_____

_____

M O D E R N   E A R T H   S C I E N C E

Chapter 14
Groundwater and Erosion

Read each statement below. If the statement is true, write _T_ in the space provided. If the statement is false, write _F_ in the space provided.

_____ 1. The water table tends to parallel the topography of the land.

_____ 2. Caverns often form when carbonic acid dissolves and enlarges cracks formed in limestone.

_____ 3. Porosity is influenced by the way sediment particles are packed together.

_____ 4. Paint pots form when water containing dissolved limestone builds a cone of rock on a cavern floor.

_____ 5. Artesian wells erupt from open pools and shoot up sheets of hot water.

_____ 6. Many desert oases are formed by artesian springs.

_____ 7. The amount of rainfall in an area has no effect on the level of the area's water table.

_____ 8. A well may go dry if the cone of depression drops to the bottom of the well.

_____ 9. A spring formed in an area with a perched water table may flow continuously.

_____ 10. Soft water contains large amounts of dissolved minerals.

Choose the one best response. Write the letter of that choice in the space provided.

_____ 11. Which of the following features would most likely be found in an area having karst topography?

    a. geysers   **b.** caverns
    c. artesian formations   **d.** travertine terraces

_____ 12. Which of the following processes is most likely to cause caverns to form?

    a. carbonation   **b.** hydration   **c.** hydrolysis   **d.** oxidation

_____ 13. The bottom region of the zone of aeration is known as the:

    a. aquifer base.   **b.** capillary fringe.
    c. cone of depression.   **d.** hydraulic gradient.

_____ 14. Karst topography occurs in regions that have experienced large amounts of:

    a. volcanic activity.   **b.** wind erosion.
    c. chemical weathering.   **d.** groundwater pollution.

Chapter 14

Choose the one best response. Write the letter of that choice in the space provided.

_____ **15.** The water table forms the upper surface of the:

    a. zone of saturation.   **b.** cap rock layer.
    c. contour surface.   **d.** artesian formation.

_____ **16.** Which of the following is the correct term for a layer of rock that is permeable and has high porosity?

    a. aquifer   **b.** travertine layer
    c. natural cavern   **d.** karst region

_____ **17.** Which of the following is most likely to occur when pesticides and fertilizers seep into soil?

    a. excessive erosion   **b.** severe drought
    c. acid rain formation   **d.** groundwater contamination

_____ **18.** Which of the following statements about groundwater is true?

    a. Groundwater is a nonrenewable resource.
    b. Groundwater can be replenished by pumping sea water into the ground.
    c. Once groundwater is polluted, it can never be used again.
    d. Groundwater can be conserved by recycling used water.

Use the diagram below to answer questions 19 and 20.

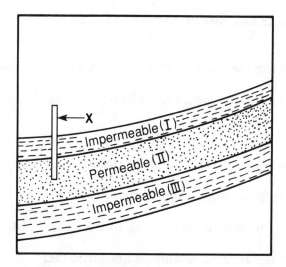

_____ **19.** The structure labeled **X** in this diagram is:

    a. a geyser.   **b.** an ordinary spring.
    c. a mud pot.   **d.** an artesian well.

_____ **20.** Rock layer **I** is called the:

    a. aquifer.   **b.** perched water table.
    c. cap rock.   **d.** zone of saturation.

Chapter 14

Complete each statement by writing the correct term or phrase in the space provided.

21. The upper surface of the zone of saturation is called the _____ .

22. Cone-shaped calcite deposits found suspended from the ceiling of a cavern are called

_____ .

23. The percentage of open space in a given volume of rock is referred to as

_____ .

24. About 90 percent of the earth's unfrozen fresh water is found in

_____ .

25. Hot springs that erupt periodically are called _____ .

26. When the water table intersects the ground surface, it forms free-flowing bodies of water

called _____ .

27. Rocks that do not allow water to flow through them are referred to as

_____ .

28. The attraction of water molecules to other materials, such as soil, is called

_____ .

29. Groundwater flows downhill in response to _____ .

30. Circular depressions formed when the roof of a cavern collapses are called

_____ .

Read each question or statement and answer it in the space provided.

31. Explain how mud pots form.

_____

_____

_____

_____

_____

HRW material copyrighted under notice appearing earlier in this work.

87

M O D E R N   E A R T H   S C I E N C E

Chapter 14

Read each question or statement and answer it in the space provided.

32. Describe ways in which groundwater may become polluted.

_____

_____

_____

_____

_____

33. Describe how the gradient of a water table affects the flow of groundwater.

_____

_____

_____

_____

_____

34. How do natural bridges form?

_____

_____

_____

_____

_____

35. A scientist discovers that a rock layer is permeable but contains very little groundwater. How can this occur?

_____

_____

_____

_____

_____

M O D E R N   E A R T H   S C I E N C E

Chapter 15
Glaciers and Erosion

Read each statement below. If the statement is true, write *T* in the space provided. If the statement is false, write *F* in the space provided.

_____ **1.** Internal plastic flow in a glacier occurs when water melted by the weight of the glacier acts as a lubricant between the glacier and the underlying rock.

_____ **2.** Ice moves faster at the surface of a glacier than at the glacier's base.

_____ **3.** The snowline is the line that marks the farthest advance of a glacier.

_____ **4.** The Great Lakes formed in what were once river valleys.

_____ **5.** The presence of glaciers in the past may be inferred from the presence of V-shaped valleys.

_____ **6.** The larger portion of an iceberg is below the surface of the water.

_____ **7.** An arête is formed by glacial deposition.

_____ **8.** Roches moutonnées are rounded knobs of bedrock produced by moving glaciers.

_____ **9.** Stratified drift is material that has been deposited by meltwater flowing from a glacier.

_____ **10.** The Milankovitch theory predicts the rate at which glaciers slide.

Choose the one best response. Write the letter of that choice in the space provided.

_____ **11.** Which of the following features form because of pressure on the surface of glaciers?

    a. crevasses   **b.** ice shelves
    c. kettles   **d.** roches moutonnees

_____ **12.** Which of the following glacial deposits is formed from stratified drift?

    a. drumlin   **b.** lateral moraine
    c. esker   **d.** medial moraine

_____ **13.** Today, a continental ice sheet is located in:

    a. Alaska.   **b.** Antarctica.   **c.** Canada.   **d.** Siberia.

_____ **14.** Which of the following is formed by the joining of arêtes?

    a. drumlin   **b.** till   **c.** esker   **d.** horn

M O D E R N   E A R T H   S C I E N C E

Chapter 15

Choose the one best response. Write the letter of that choice in the space provided.

_____ **15.** When a tributary glacier melts, it may leave a suspended formation called:

        a. a hanging valley   **b.** a snowline.

        c. a kettle   **d.** an erratic.

_____ **16.** Firn is formed as a result of:

        a. the polishing of bedrock by the scraping action of rock particles embedded in glacial ice.

        b. changes in the flow rate of glacial meltwater.

        c. the partial melting and refreezing of snow crystals into small grains of ice.

        d. the deposition of unsorted glacial drift.

_____ **17.** Which of the following is the most commonly accepted cause of ice ages?

        a. changes in the amount of heat produced by the sun

        b. blockage of the sun's rays by volcanic dust

        c. movements of continents affecting warm ocean currents

        d. small changes in the earth's orbit, tilt, and precession

Use the diagram below to answer questions 18-20.

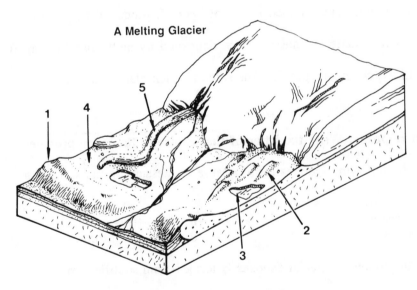

A Melting Glacier

_____ **18.** Which feature in the diagram represents a terminal moraine?

    a. 1   **b.** 2   **c.** 3   **d.** 4

_____ **19.** Which feature in the diagram may be formed by streams of meltwater flowing in tunnels beneath glacial ice?

    a. 2   **b.** 3   **c.** 4   **d.** 5

_____ **20.** Which feature in the diagram is formed from stratified drift?

    a. 1   **b.** 2   **c.** 4   **d.** 5

M O D E R N   E A R T H   S C I E N C E

Chapter 15

Complete each statement by writing the correct term or phrase in the space provided.

21. Travel over the top of a glacier is dangerous due to the presence of deep cracks called

_____ .

22. Scientists have discovered that the temperature of the water in which marine animals live is related

to the ratio of two isotopes of _____ in the animals' shells.

23. A long period of climatic cooling during which ice sheets cover large areas of the earth's

surface is known as an _____ .

24. An almost motionless mass of permanent snow and ice is called a

_____ .

25. The type of glacier that generally forms in mountainous areas is called a

_____ .

26. Lake Winnipeg in Canada is the chief remnant of the former glacial Lake

_____ .

27. Glacial movement resulting from the slippage of ice crystals over each other is called

_____ .

28. A bowl-shaped depression eroded by a valley glacier is called a _____ .

29. Deposits of stratified drift that form in front of a glacier's terminal moraine are called

_____ .

30. The composition of water in a salt lake depends on the nature of dissolved

_____ .

Read each question or statement and answer it in the space provided.

31. Explain how a snowfield becomes a glacier.

_____

_____

_____

_____

M O D E R N   E A R T H   S C I E N C E

Chapter 15

Read each question or statement and answer it in the space provided.

32. How does a glacier move by basal slip?

_____

_____

_____

_____

33. Describe the formation of a salt lake.

_____

_____

_____

_____

34. What are three kinds of periodic changes that occur in the way the earth moves around the sun?

_____

_____

_____

_____

35. How would an increase in the amount of solar energy reaching the earth's surface most likely affect sea level along the coastline of the United States?

_____

_____

_____

_____

M O D E R N   E A R T H   S C I E N C E

Chapter 16
Erosion by Wind and Waves

Read each statement below. If the statement is true, write _T_ in the space provided. If the statement is false, write _F_ in the space provided.

_____ 1. A ventifact's appearance indicates the direction of the wind that formed it.

_____ 2. The windward side of a dune has a steeper slope than the slipface.

_____ 3. Sea cliffs usually erode evenly.

_____ 4. Storms cause barrier islands to migrate toward the shoreline.

_____ 5. The berm of a beach is less steep in winter than in summer.

_____ 6. A longshore current will transport sand in a direction parallel to the shoreline.

_____ 7. Sea level is currently rising due to melting glaciers.

_____ 8. Scientists believe that the side of a continent near the trailing edge of a moving crustal plate tends to be uplifted.

_____ 9. An emergent coastline is characterized by the presence of few bays or headlands and many long, wide beaches.

_____ 10. Changes in exposed coastline can be caused by the rising or sinking of land as well as by changes in sea level.

Choose the one best response. Write the letter of that choice in the space provided.

_____ 11. What are most sand grains made of?

        a. coral   **b.** quartz   **c.** dust   **d.** loess

_____ 12. The closely packed, small rocks left behind by deflation are called:

        a. deposits.   **b.** ventifacts.
        c. desert pavement.   **d.** sedimentary rock.

_____ 13. What type of dune forms at right angles to the wind?

        a. barchan   **b.** longitudinal   **c.** transverse   **d.** parabolic

_____ 14. A sea cliff composed of which of the following would probably erode most slowly?

        a. glacial deposits   **b.** sedimentary rock
        c. granite   **d.** loess

M O D E R N   E A R T H   S C I E N C E

Chapter 16

Choose the one best response. Write the letter of that choice in the space provided.

_____ 15. Which of the following is the result of the greatest amount of wave erosion?

      a. sea arch   **b.** sea cave   **c.** sea cliff   **d.** sea stack

_____ 16. Sand that is carried by waves from the berm and deposited offshore usually forms a:

      a. sand bar.   **b.** longitudinal dune.
      c. spit.   **d.** cobble.

_____ 17. The coastal glacial valleys that become flooded as sea level rises are called:

      a. estuaries.   **b.** fiords.   **c.** headlands.   **d.** berms.

_____ 18. Approximately what percentage of the land in the United States is coastal?

      a. 5%   **b.** 10%   **c.** 15%   **d.** 20%

Use the diagram below to answer questions 19 and 20.

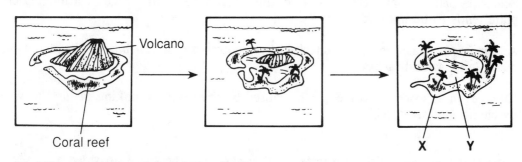

_____ 19. What is the area labeled **X** called?

      a. a barrier island   **b.** a fringing reef
      c. a spit   **d.** an atoll

_____ 20. What is the area labeled **Y** called?

      a. a lagoon   **b.** an estuary
      c. a tidal flat   **d.** a submergent coastline

Complete each statement by writing the correct term or phrase in the space provided.

21. The rolling and bouncing movement of sand grains being blown by wind is called

_____ .

22. The process in which wind removes the top layer of fine particles from soil is called

_____ .

M O D E R N   E A R T H   S C I E N C E

Chapter 16

Complete each statement by writing the correct term or phrase in the space provided.

23. The shallow depression left after dry, bare soil is exposed by wind erosion is called a

_____ .

24. The accumulation of dust into thick deposits of fine, yellowish sediment results in the

formation of _____ .

25. A fairly level platform of rock that lies beneath the water at the base of a sea cliff is called a

_____ .

26. On beaches, the raised midsections formed by sand deposits are called

_____ .

27. A resistant rock area that projects out from the shore is called a

_____ .

28. Narrow regions of shallow water between barrier islands and shorelines are called

_____ .

29. The type of dune that generally forms when sand is blown from a deflation hollow and

collects around its rim is called a _____ .

30. Waves often cut into fractures and weak areas along the base of a cliff, forming a large hole

called a _____ .

Read each question or statement and answer it in the space provided.

31. Explain how the formation of a wave-built terrace helps to slow the rate of sea cliff erosion.

_____

_____

_____

_____

M O D E R N   E A R T H   S C I E N C E

Chapter 16

Read each question or statement and answer it in the space provided.

32. How do waves promote the process of chemical weathering of a shoreline?

_____

_____

_____

_____

33. What are three factors that have allowed high tides to flood Venice, and how did these factors contribute to the problem?

_____

_____

_____

_____

34. How do waves form beaches?

_____

_____

_____

_____

35. What types of material would most likely compose the beach of a volcanic island that is surrounded by a barrier reef? Why?

_____

_____

_____

_____

M O D E R N   E A R T H   S C I E N C E

Unit 4
Reshaping the Crust

**Read each statement below. If the statement is true, write _T_ in the space provided.
If the statement is false, write _F_ in the space provided.**

_____ **1.** Hanging valleys are caused by glacial meltwater.

_____ **2.** Basal slip occurs when the weight of a glacier exerts enough pressure to melt the ice where it touches the ground, forming a lubricant.

_____ **3.** The Milankovitch theory states that ice ages are caused when volcanic dust decreases the amount of sunlight reaching the earth's surface.

_____ **4.** Weathering of rocks occurs slowly in hot, humid climates.

_____ **5.** Galvanized steel resists rusting.

_____ **6.** Waves striking the shore are sometimes powerful enough to be recorded on seismographs.

_____ **7.** Sand bars are formed when sand carried away from the beach is deposited offshore.

_____ **8.** A local water budget is always balanced.

_____ **9.** Erosion often increases the amount of land drained in a watershed.

_____ **10.** Alluvial fans are also called river deltas.

_____ **11.** Due to recent conservation efforts, coastlines in the United States are no longer in danger of being damaged.

_____ **12.** The surface of a glacier flows faster than its base.

_____ **13.** Rocks that do not allow water to pass through them easily are called impermeable.

_____ **14.** A geyser is a common feature in areas of karst topography.

_____ **15.** A prerequisite for an artesian formation is that the sloping permeable layer must be exposed to the earth's surface.

MODERN EARTH SCIENCE

Unit 4

Choose the one best response. Write the letter of that choice in the space provided.

_____ **16.** A muddy or sandy part of the shoreline that is visible at low tide and submerged at high tide is a:

    a. fringing reef.   **b.** barrier island.
    c. tidal flat.   **d.** deflation hollow.

_____ **17.** A flat landform that lies on top of horizontal rock layers near sea level is called a:

    a. peneplain.   **b.** plateau.   **c.** butte.   **d.** plain.

_____ **18.** Uncontaminated rainwater is naturally acidic due to the presence of:

    a. silica.   **b.** carbon dioxide.
    c. oxygen.   **d.** iron oxide.

_____ **19.** Roches moutonnees are best described as:

    a. long, low mounds of till.
    b. ridges of gravel and coarse sand.
    c. rounded rock projections.
    d. groups of loose boulders.

_____ **20.** Which of the following has the smallest particle size?

    a. clay   **b.** silt   **c.** sand   **d.** gravel

_____ **21.** In which of the following climates is solifluction most likely to occur?

    a. tropical   **b.** desert   **c.** temperate   **d.** arctic

_____ **22.** A ground moraine is formed primarily from till deposited:

    a. at the glacier's leading edge.
    b. along the glacier's sides.
    c. beneath a glacier.
    d. at the glacier's source region.

_____ **23.** Evidence for the cause of ice ages has been found in the

    a. mountains   **b.** atmosphere   **c.** valleys   **d.** ocean

_____ **24.** Condensation is a change in water from a:

    a. solid to liquid.   **b.** liquid to solid.
    c. gas to liquid.   **d.** liquid to gas.

_____ **25.** A wind gap is a water gap that has been:

    a. deepened.   **b.** lengthened.   **c.** abandoned.   **d.** eroded.

M O D E R N E A R T H S C I E N C E

Unit 4

Choose the one best response. Write the letter of that choice in the space provided.

_____ **26.** Glacial meltwater usually appears milky due to the presence in the water of:

    a. fine rock particles.   **b.** ice crystals.
    c. air bubbles.   **d.** dissolved salts.

_____ **27.** Most sinkholes are formed by:

    a. collapse of a cavern roof.   **b.** volcanic activity.
    c. stream erosion.   **d.** travertine deposits.

_____ **28.** Karst topography is a result of:

    a. geothermal activity.   **b.** groundwater contamination.
    c. chemical weathering.   **d.** stream erosion.

_____ **29.** A mound of wind-blown sand is called a:

    a. loess.   **b.** dune.   **c.** pavement.   **d.** spit.

Complete each statement by writing the correct term or phrase in the space provided.

30. Ice ages probably begin with a drop in average temperatures combined with an increase in

_____ .

31. The groundwater layer in which all pore spaces are filled with water is called the

_____ .

Read each question and answer it in the space provided.

32. What is the name for the process by which the products of weathering are transported?   _____

33. What is the term for sediment deposited by a glacier?   _____

34. What is the name for a structure that periodically erupts with steam and water?   _____

35. The removal of which soil horizon would probably lead to the greatest reduction of soil fertility?   _____

36. What mineral are most sand grains made of?   _____

37. What is the term for raised river banks produced by the accumulation of deposits?   _____

M O D E R N   E A R T H   S C I E N C E

Unit 4

Read each question and answer it in the space provided.

38. Which side of a dune generally has the gentlest slope? _____

39. What type of stream load consists of sand, silt, and mud? _____

40. What are chains of sand ridges that connect islands called? _____

41. In what stage of a river do large meanders develop? _____

42. What has been the main cause of sea level rise over the last 11,000 years? _____

43. What is the term for the volume of water moved by a river in a given time? _____

44. What is the term for a long, narrow offshore ridge of sand that may be up to 100 km long and lies nearly parallel to the shoreline? _____

Use the diagram below to answer questions 45 and 46.

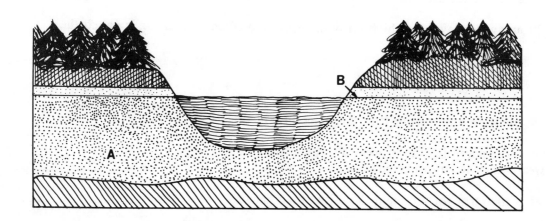

45. What is the term for water in the area labeled **A**? _____

46. What is the term for the level indicated by label **B**? _____

M O D E R N   E A R T H   S C I E N C E

Unit 4

Read each question or statement and answer it in the space provided.

47. How can a scientist tell if a glacier has passed through a particular valley?

_____

_____

_____

_____

_____

_____

48. Describe how plants and animals contribute to the weathering of rocks.

_____

_____

_____

_____

_____

_____

49. Describe the formation of an atoll.

_____

_____

_____

_____

_____

_____

M O D E R N   E A R T H   S C I E N C E

Unit 4

Read each question or statement and answer it in the space provided.

50. Describe the major factors affecting local water budgets.

_____

_____

_____

_____

_____

_____

MODERN EARTH SCIENCE

Chapter 17
The Rock Record

Read each statement below. If the statement is true, write T in the space provided. If the statement is false, write F in the space provided.

_____ 1. The earth is believed to be about 4.6 billion years old.

_____ 2. Ripple marks can be used to determine the original arrangement of rock layers in an area.

_____ 3. Erosion rates have been used to establish the age of the Grand Canyon accurately.

_____ 4. One varve represents two years of deposition of sedimentary particles.

_____ 5. Scientists are determining the age of minerals using laser beams to discover when the oldest rocks on earth were formed.

_____ 6. The principle of uniformitarianism states that all the earth's geologic features formed at the same time.

_____ 7. Fossil spores and pollen indicate that Antarctica had a much warmer climate 40 million years ago.

_____ 8. Mold fossils generally provide little information about the internal structures of the original organisms.

_____ 9. The hard parts of an organism, such as bones and teeth, are usually the only parts of the body to be preserved in the rock record.

_____ 10. A footprint is an example of a trace fossil.

Choose the one best response. Write the letter of that choice in the space provided.

_____ 11. A fossil animal with its skin and muscles intact was most likely preserved through the process of:

     a. petrification.   **b.** freezing.
     c. carbonization.   **d.** casting.

_____ 12. Sedimentary deposits that form in glacial lakes and show definite annual layers are called:

     a. varves.   **b.** dikes.   **c.** sills.   **d.** crossbeds.

_____ 13. Which process below forms fossils by drying the organic matter in an organism's body?

     a. mummification   **b.** petrification
     c. excretion   **d.** deposition

M O D E R N E A R T H S C I E N C E

Chapter 17

Choose the one best response. Write the letter of that choice in the space provided.

_____ **14.** Which type of unconformity is the most difficult to identify by sight in rock strata?

    a. nonconformity   **b.** angular unconformity
    c. disconformity   **d.** crossbed intrusion

_____ **15.** Which of the following principles would most likely be used to determine the relative age of undisturbed sedimentary strata?

    a. principle of uniformitarianism
    b. law of superposition
    c. principle of intrusion
    d. law of crosscutting relationships

Use the diagram below to answer questions 16–18.

Diagram of Undisturbed Strata

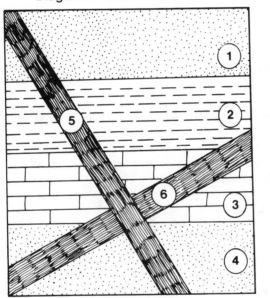

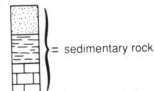
= igneous rock

= sedimentary rock

_____ **16.** Which rock layer in the diagram contains the oldest rock?

    a. 1   **b.** 2   **c.** 3   **d.** 4

_____ **17.** Which of the following would be used to determine the difference in the relative ages of the rock in layer **6** and layer **3**?

    a. law of superposition
    b. principle of average erosion rates
    c. varve counts
    d. law of crosscutting relationships

_____ **18.** Which feature in the diagram contains the youngest rock?

    a. 1   **b.** 4   **c.** 5   **d.** 6

M O D E R N   E A R T H   S C I E N C E

Chapter 17

Choose the one best response. Write the letter of that choice in the space provided.

_____ 19. A nonerosional boundary between two layers of sedimentary rock is called a:

      a. foliation band.   **b.** stratification barrier.
      c. disconformity line.   **d.** bedding plane.

_____ 20. Which of the following elements is most appropriate for dating a fossil between 8,000 and 20,000 years old?

      a. uranium-238   **b.** thorium-234   **c.** potassium-40   **d.** carbon-14

Complete each statement by writing the correct term or phrase in the space provided.

21. The principle that states that current geologic processes are the same processes that occurred in the past is known as the principle of _____ .

22. The period of time in which half of a radioactive sample decays is called the

_____ .

23. An unconformity that consists of stratified sedimentary rock above a horizontal basalt surface would be classified as a type of _____ .

24. The final product formed by a series of radioactive decays is called the original element's

_____ .

25. Scientists who study fossils to learn about the earth's geologic history are called

_____ .

26. Trace fossils are usually found in rocks that form through the process of

_____ .

27. Fossils formed when mud fills in a mold and hardens are called _____ .

28. Smooth, rounded stones sometimes found within the remains of a dinosaur are called

_____ .

29. Fossils such as trilobites that can be used to establish the age of a rock are called

_____ .

30. The change in living things over time is called the process of _____ .

M O D E R N   E A R T H   S C I E N C E

Chapter 17

Read each question or statement and answer it in the space provided.

31. What are the characteristics of a good index fossil?

_____

_____

_____

32. How can scientists use the average rate of erosion to determine the age of features such as streams? What are the limitations of this method?

_____

_____

_____

33. Describe the process of petrification.

_____

_____

_____

34. Describe the process of radioactive decay.

_____

_____

_____

35. How do geologists use the law of superposition and the law of crosscutting relationships to determine the relative ages of faulted and eroded rock layers?

_____

_____

_____

M O D E R N   E A R T H   S C I E N C E

Chapter 18
A View of the Earth's Past

Read each statement below. If the statement is true, write _T_ in the space provided. If the statement is false, write _F_ in the space provided.

_____ 1. The presence of stromatolite fossils in Precambrian rocks indicates that shallow seas may have covered much of the earth during some periods of Precambrian time.

_____ 2. Fossil bacteria and algae, thought to be the earth's first life forms, have been found in rocks 4.5 billion years old.

_____ 3. The theory of evolution states that organisms change over time and new kinds of organisms are derived from ancestral types.

_____ 4. By the end of the Paleozoic Era, the earth's landmasses were grouped together to form the supercontinent Pangaea.

_____ 5. Land plants first developed during the Silurian Period.

_____ 6. The Mesozoic Era is known as the Age of Fishes.

_____ 7. The end of the Jurassic Period was marked by mass extinctions.

_____ 8. Fossils of the ancestors of modern humans date to almost 2 million years ago.

_____ 9. Brachiopods, a group of shelled animals, are characteristic of the Cambrian Period.

_____ 10. The earth's crust solidified during the Cambrian Period.

Choose the one best response. Write the letter of that choice in the space provided.

_____ 11. What do scientists now often use to determine the relative age of a rock layer?

    a. mineral content   **b.** the geologic column
    c. rock type   **d.** the geologic calendar

_____ 12. The current geologic era is the:

    a. Mesozoic.   **b.** Precambrian.   **c.** Paleozoic.   **d.** Cenozoic.

_____ 13. Some eras are divided into time units characterized by specific fossils and called:

    a. epochs.   **b.** events.   **c.** periods.   **d.** eons.

_____ 14. Natural selection is also known as survival of the:

    a. largest.   **b.** young.   **c.** strongest.   **d.** fittest.

HRW material copyrighted under notice appearing earlier in this work.

107

M O D E R N   E A R T H   S C I E N C E

Chapter 18

Choose the one best response. Write the letter of that choice in the space provided.

_____ **15.** Which of the following are rare in Precambrian rocks?

    a. fossil remains   **b.** sandstones
    c. minerals   **d.** metamorphisms

_____ **16.** Large areas of exposed Precambrian rocks are called:

    a. mesas.   **b.** basins.   **c.** shields.   **d.** shelves.

_____ **17.** About how many million years ago did the Paleozoic Era begin?

    a. 24   **b.** 144   **c.** 438   **d.** 570

Use the table below to answer questions 18-20.

EPOCH	TIME (millions of years ago)	CHARACTERISTIC ORGANISMS
1. Pliocene	5	carnivores, first modern horses
2. Miocene	24	species of deer, rhinoceros, pig, saber-toothed cat
3. Oligocene	37	species of horses, cats, dogs
4.	58	flying squirrels and bats, whales
5. Paleocene		rodents, lemuroids

_____ **18.** The missing epoch that belongs in the space labeled **4** is called the:

    a. Jurassic.   **b.** Pleistocene.   **c.** Cretaceous.   **d.** Eocene.

_____ **19.** During which epoch did the largest known land mammal, called Baluchitherium, live?

    a. Pliocene   **b.** Miocene   **c.** Eocene   **d.** Paleocene

_____ **20.** About how many million years ago did the Paleocene epoch begin?

    a. 245   **b.** 144   **c.** 65   **d.** 58

Read each question and answer it in the space provided.

21. In what period did vertebrates first appear?   _____

22. What is the general term for the method that allows more accurate dating of absolute ages of rock layers than was previously possible based on fossils and rates of deposition?   _____

23. Who proposed the theory of evolution by natural selection?   _____

M O D E R N   E A R T H   S C I E N C E

Chapter 18

Read each question and answer it in the space provided.

24. What are the two periods of the Cenozoic Era? _____

25. During what era did birds first appear? _____

26. What epoch ended with the end of the most recent ice age? _____

27. What were the most common of the Cambrian invertebrates called? _____

28. The presence of a limestone layer indicates an area that was once covered by what? _____

29. Scientists have found evidence suggesting that mass extinctions of organisms occurred at about the same time as the earth experienced what dramatic cosmic event? _____

30. The extinction of dinosaurs marked the end of what period? _____

Read each question or statement and answer it in the space provided.

31. Why can the Cenozoic Era be easily divided into numerous smaller time units?

_____

_____

_____

_____

_____

_____

32. Use the theory of evolution by natural selection to explain the extinction of some organisms during times of climatic change.

_____

_____

_____

_____

_____

M O D E R N   E A R T H   S C I E N C E

Chapter 18

Read each question or statement and answer it in the space provided.

33. Why is the Precambrian rock record so difficult to interpret?

_____

_____

_____

_____

_____

_____

34. What does "Carboniferous" mean? Why is this time unit so named?

_____

_____

_____

_____

_____

_____

35. How did climatic conditions in the Mesozoic Era favor reptile survival and development?

_____

_____

_____

_____

_____

_____

M O D E R N   E A R T H   S C I E N C E

Chapter 19
The History of the Continents

Read each statement below. If the statement is true, write T in the space provided. If the statement is false, write F in the space provided.

_____ 1. The northward movement of Laurasia separated North America from Africa.

_____ 2. The lithosphere is made up of plates that ride on the asthenosphere.

_____ 3. Outlines of the continents as they existed in Precambrian time have been determined by fossil evidence.

_____ 4. Pangaea was initially located near the North Pole.

_____ 5. The Sahara was once covered by a sheet of ice.

_____ 6. The modern Mediterranean Sea formed when the east end of the Tethys Sea was closed off.

_____ 7. Australia was once part of Gondwanaland.

_____ 8. The Canadian Shield is an area of exposed Precambrian rock.

_____ 9. Salt deposits in Kansas are thought to have formed when a large inland sea dried up.

_____ 10. The Appalachian Mountains existed during Precambrian time.

Choose the one best response. Write the letter of that choice in the space provided.

_____ 11. A geosyncline is defined as:

    a. a large inland sea.
    b. a sediment-filled trough.
    c. an outcrop of Precambrian rock.
    d. an uplifted plateau.

_____ 12. Approximately how old are fossils found in Precambrian rocks of North America?

    a. 2 million years old   **b.** 200 million years old
    c. 1 billion years old   **d.** 2 billion years old

_____ 13. Which of the following was once part of Laurasia?

    a. South America   **b.** Antarctica
    c. North America   **d.** Africa

_____ 14. New pieces of continental crust are called:

    a. cratons.   **b.** shields.   **c.** plateaus.   **d.** terranes.

M O D E R N   E A R T H   S C I E N C E

Chapter 19

Choose the one best response. Write the letter of that choice in the space provided.

_____ 15. What part of North America has remained mostly intact for the last 600 million years?

      a. Canadian Shield   **b.** Basin and Range Province
      c. Rocky Mountain area   **d.** Colorado Plateau

_____ 16. If the continents continue to move as predicted, in 150 million years Africa will collide with:

      a. South America.   **b.** Eurasia.
      c. India.   **d.** North America.

_____ 17. The Mississippian Period is represented in the Grand Canyon by the:

      a. Redwall limestone.   **b.** Coconino sandstone.
      c. Kaibab limestone.   **d.** Tapeats sandstone.

_____ 18. During what geologic period was the Bright Angel shale deposited?

      a. Cambrian   **b.** Ordovician   **c.** Silurian   **d.** Devonian

Use the table below to answer questions 19 and 20.

	Features of the Paleozoic Earth
1	the single large ocean covering 60% of the earth's surface
2	the triangular body of water that cut into the eastern edge of the supercontinent
3	the northern segment of the giant continent
4	the southern segment of the giant continent
5	the large continental mass covering 40% of the earth's surface

_____ 19. What number in the table represents Panthalassa?

      a. 1   **b.** 2
      c. 3   **d.** 4

_____ 20. What number in the table represents Gondwanaland?

      a. 2   **b.** 3
      c. 4   **d.** 5

Read each question and answer it in the space provided.

21. By the beginning of which geologic era had North America begun to take on the shape it has today? _____

22. What is the correct term for any type of exposed rock formation? _____

23. What geologic law assumes that young geologic deposits sit on top of older deposits? _____

Read each question and answer it in the space provided.

Chapter 19

Read each question and answer it in the space provided.

24. What body of water was formed by the opening of a rift between Africa and South America during the Cretaceous Period?

_____

25. What method is used to estimate the absolute age of fossils?

_____

26. What features most likely caused cross-bedding seen in Grand Canyon sandstones?

_____

27. During what geologic era did the Rocky Mountains begin to form?

_____

28. In North America, Precambrian rocks contain fossils of what type of organisms?

_____

29. Most of the major landforms of modern North America were created during what era?

_____

30. What is the name of the uplifted area east of the Sierra Nevada through which the Colorado River flows?

_____

Read each question or statement and answer it in the space provided.

31. Describe the configuration of the west coast of North America that scientists predict will exist 150 million years from now.

_____

_____

_____

_____

_____

32. Describe how the first humans may have reached North America.

_____

_____

_____

_____

_____

M O D E R N   E A R T H   S C I E N C E

Chapter 19

Read each question or statement and answer it in the space provided.

33. What would most likely happen if the Grand Canyon area were uplifted again?

_____

_____

_____

_____

_____

34. What sequence of events can be traced by studying fossils in successive rock layers of the Grand Canyon?

_____

_____

_____

_____

_____

35. Describe how the Grand Canyon formed.

_____

_____

_____

_____

_____

M O D E R N   E A R T H   S C I E N C E

Unit 5
The History of the Earth

**Read each statement below. If the statement is true, write _T_ in the space provided.
If the statement is false, write _F_ in the space provided.**

_____ 1. A species that lived during several different geologic time periods would make a good index fossil.

_____ 2. Australia and Antarctica were once part of the same landmass.

_____ 3. The geologic column allows scientists to estimate the age of rock layers that contain no radioactive minerals.

_____ 4. There is evidence that the layers of sedimentary rock in the Grand Canyon have been overturned.

_____ 5. Sandstone layers in the Grand Canyon often contain fossils of marine organisms.

_____ 6. The laws of superposition and crosscutting relationships could be used to determine the relative age of an unconformity.

_____ 7. The final product of radioactive decay is usually a nonradioactive element.

_____ 8. The Cenozoic Era was a time of increased tectonic activity.

_____ 9. The part of the supercontinent called Laurasia was located north of the Tethys Sea.

_____ 10. Changes in organisms can sometimes be traced by examining their remains in successive rock layers.

_____ 11. Geologists use major changes in the earth's surface or climate and extinctions of various species to divide the geologic time scale into smaller units.

_____ 12. Precambrian rocks contain abundant fossils.

_____ 13. South America and Africa were part of the same landmass that broke from Gondwanaland.

_____ 14. A fault is always older than the rocks through which it cuts.

_____ 15. In an angular unconformity, the bedding planes of older rock layers are not parallel to the bedding planes of younger layers.

_____ 16. Mammals were the dominant animals of the Paleozoic Era.

M O D E R N   E A R T H   S C I E N C E

Unit 5

Choose the one best response. Write the letter of that choice in the space provided.

_____ **17.** In North America, volcanic activity during the Cenozoic Era occurred chiefly in the:

    a. Northwest.   **b.** Southwest.   **c.** Northeast.   **d.** Southeast.

_____ **18.** Which of the following would most likely be used to determine the relative age of a fault?

    a. law of superposition
    b. radioactive decay
    c. principle of uniformitarianism
    d. law of crosscutting relationships

_____ **19.** What is the second epoch of the Quaternary Period called?

    a. Miocene   **b.** Permian   **c.** Mississippian   **d.** Holocene

_____ **20.** Geologic evidence shows that the first land plants began to grow during which era?

    a. Precambrian   **b.** Cenozoic   **c.** Paleozoic   **d.** Mesozoic

_____ **21.** A rock contains 4 milligrams of a radioactive element. At the end of 60 years, only 0.5 milligram of the element remains in the rock. What is the half-life of the element?

    a. 60 years   **b.** 30 years   **c.** 20 years   **d.** 15 years

_____ **22.** Which type of fossil would yield the most information about the internal structure of an organism?

    a. a tree imprint   **b.** a tree cast
    c. a tree mold   **d.** a petrified tree

_____ **23.** Environmental conditions during the Mesozoic Era especially favored the survival of:

    a. reptiles.   **b.** woolly mammoths.
    c. birds.   **d.** early humans.

_____ **24.** From which period have rock layers in the Grand Canyon probably been completely eroded?

    a. Cambrian   **b.** Silurian   **c.** Mississippian   **d.** Permian

_____ **25.** About 150 million years from now, the portion of California west of the San Andreas fault will probably be located near what present-day region?

    a. Mexico   **b.** Hawaii   **c.** Oregon   **d.** Alaska

M O D E R N   E A R T H   S C I E N C E

Unit 5

Choose the one best response. Write the letter of that choice in the space provided.

_____ **26.** During the Pleistocene Epoch, a land bridge connected North America and:

    a. Africa.   **b.** Australia.   **c.** Greenland.   **d.** Eurasia.

_____ **27.** Which layer of the Grand Canyon was deposited during the Cambrian Period?

    a. Hermit shale   **b.** Kaibab limestone
    c. Bright Angel shale   **d.** Redwall limestone

Use the table below to answer question 28.

Period	Years Before Present
	65,000,000
Cretaceous	
	144,000,000
Jurassic	
	208,000,000
Triassic	
	245,000,000

_____ **28.** What geologic era is represented by this table?

    a. Precambrian   **b.** Paleozoic   **c.** Mesozoic   **d.** Cenozoic

_____ **29.** The principle of uniformitarianism states that:

    a. rocks can be dated by the radioactive elements they contain.
    b. geologic processes that occurred in the past are still at work today.
    c. rock layers that are buried are older than the layers above them.
    d. the major geologic features of the earth all formed at the same time.

_____ **30.** A period of erosion shows up in layers of rock as:

    a. an unconformity.   **b.** a varve.
    c. a fault.   **d.** an intrusion.

Read each question and answer it in the space provided.

31. During which geologic era did Pangaea exist?   _____

32. Where in the geologic column are the most recent rocks
located?   _____

33. What part of Pangaea was located south of the Tethys
Sea?   _____

MODERN EARTH SCIENCE

Unit 5

Read each question and answer it in the space provided.

34. What were the most common invertebrates of the
Cambrian Period?

_____

35. What is the correct term for the type of hardened tree
sap in which insects are occasionally preserved?

_____

36. What landmass split to form North America and Eurasia?

_____

37. What is the name of the exposed portion of the craton
around which North America has been built up?

_____

38. What theory is supported by evidence that organisms
change over time and that new kinds of organisms are
derived from ancestral types?

_____

39. What ocean covered about 60 percent of the earth's
surface when Pangaea existed?

_____

40. Which geologic time unit is usually named for the
location in which rocks containing identifying fossils were
first found?

_____

41. What is the correct term for fossilized dung or animal
waste material?

_____

42. What type of fossil is a dinosaur footprint?

_____

43. What is the correct term for the process by which the
original organic material in an organism is replaced by
minerals?

_____

44. During times of major geologic or climactic changes,
what happens to organisms that cannot adapt to the
changes?

_____

45. Rocks of what era contain nearly half of the world's
deposits of valuable minerals?

_____

M O D E R N   E A R T H   S C I E N C E

Unit 5

Read each question or statement and answer it in the space provided.

46. What geologic features would indicate that several sedimentary rock layers had been overturned?

_____

_____

_____

_____

_____

_____

47. What effect did northward drift toward the equator have on Pangaea?

_____

_____

_____

_____

_____

48. How does the arrival of ancestors of modern humans in the Pleistocene Epoch help to explain the extinction of some animal species at that time?

_____

_____

_____

_____

_____

M O D E R N   E A R T H   S C I E N C E

Unit 5

Read each question or statement and answer it in the space provided.

49. How and when did the Alps and Himalayas form?

_____

_____

_____

_____

_____

_____

50. Explain how scientists use carbon dating to determine the age of organic samples.

_____

_____

_____

_____

_____

_____

M O D E R N E A R T H S C I E N C E

Chapter 20
The Ocean Basins

Read each statement below. If the statement is true, write *T* **in the space provided. If the statement is false, write** *F* **in the space provided.**

_____ **1.** Many of the features discovered on the ocean floor were first detected by sonar.

_____ **2.** Most trenches are located within the mid-ocean ridge system.

_____ **3.** Most material found in oozes on the deep-sea bottom is derived from the continents.

_____ **4.** The world's oceans contain more than 97 percent of the earth's mass.

_____ **5.** The global ocean is divided into three principal oceans.

_____ **6.** Most siliceous ooze is found around the continent of Antarctica.

_____ **7.** Underwater continental crust makes up part of the deep ocean basin.

_____ **8.** The shoreline marks the boundary between the continent and the ocean floor.

_____ **9.** The coarser sediments of the continental margin are usually found close to shore.

_____ **10.** The width of the continental shelf varies depending on its location.

Choose the one best response. Write the letter of that choice in the space provided.

_____ **11.** Approximately what percentage of the earth's surface is covered by oceans?

    a. 10%   **b.** 35%   **c.** 70%   **d.** 90%

_____ **12.** Submarine canyons most likely formed due to erosion by:

    a. turbidity currents.   **b.** volcanic activity.
    c. transform faulting.   **d.** trench formation.

_____ **13.** Scientists obtain deep-ocean core samples in order to analyze:

    a. deep-ocean basin currents.
    b. composition of ocean floor sediments.
    c. deep-ocean temperatures.
    d. composition of seawater.

_____ **14.** Which of the following is a major source of calcium carbonate in organic sediments on the ocean floor?

    a. radiolaria   **b.** diatoms   **c.** foraminifera   **d.** nodules

M O D E R N   E A R T H   S C I E N C E

Chapter 20

Choose the one best response. Write the letter of that choice in the space provided.

_____ **15.** The greatest ocean depths are found within:

      a. trenches.   **b.** continental slopes.
      c. abyssal plains.   **d.** continental margins.

_____ **16.** The difference between a bathysphere and a bathyscaph is that a bathysphere:

      a. carries passengers.   **b.** remains connected to a ship.
      c. uses sonar.   **d.** is self-propelled.

Use the diagram below to answer questions 17 and 18.

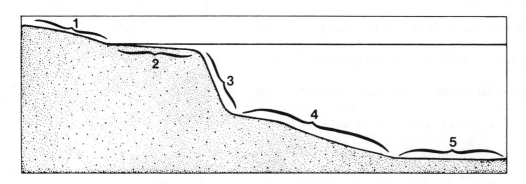

_____ **17.** Which region in this diagram represents the continental shelf?

      a. 1   **b.** 2   **c.** 3   **d.** 4

_____ **18.** Which region is composed of oceanic crust?

      a. 2   **b.** 3   **c.** 4   **d.** 5

_____ **19.** The deepest of the earth's principal oceans is the:

      a. Arctic.   **b.** Pacific.   **c.** Atlantic.   **d.** Indian.

_____ **20.** Which of the following is largely composed of diatoms?

      a. red ooze   **b.** calcareous ooze
      c. siliceous ooze   **d.** lava

Complete each statement by writing the correct term or phrase in the space provided.

21. The raised wedge of sediments at the base of a continental slope is called the

_____ .

22. A small area of ocean that is partially surrounded by land is called a

_____ .

M O D E R N   E A R T H   S C I E N C E

Chapter 20

Complete each statement by writing the correct term or phrase in the space provided.

23. One common type of mud on the ocean floor is red _____ .

24. Isolated volcanic mountains scattered on the ocean floor are called

    _____ .

25. The Gulf Stream and the Labrador Current meet at the fishing grounds called the

    _____ .

26. The deep, V-shaped valleys found on the continental slope are called

    _____ .

27. The extremely flat areas in the deep ocean are the _____ .

28. Turbidity currents appear to be the result of _____ .

29. Some portions of the ocean floor are covered with potato-shaped lumps of minerals called

    _____ .

30. The branch of earth science that includes the study of the physical characteristics of the ocean

    is called _____ .

Read each question or statement and answer it in the space provided.

31. Why is sea level lower during glacial periods than it is now?

    _____

    _____

    _____

    _____

    _____

32. Describe how guyots are formed.

    _____

    _____

    _____

    _____

    _____

MODERN EARTH SCIENCE

Chapter 20

Read each question or statement and answer it in the space provided.

33. How do icebergs provide ocean-floor sediments?

_____

_____

_____

_____

_____

34. How is the depth of water determined by using sonar?

_____

_____

_____

_____

_____

35. How might the number of fish and shellfish be different in the Grand Banks area if it had a very narrow continental shelf?

_____

_____

_____

_____

_____

M O D E R N   E A R T H   S C I E N C E

Chapter 21
Ocean Water

Read each statement below. If the statement is true, write _T_ in the space provided. If the statement is false, write _F_ in the space provided.

_____ 1. Organic remains in ocean water are broken down by bacteria.

_____ 2. The Arctic Ocean is covered by pack ice during most of the year.

_____ 3. Pure fresh water is denser than ocean water.

_____ 4. The temperature of ocean water drops sharply not far below the ocean surface.

_____ 5. Zooplankton use the energy from sunlight to carry on photosynthesis.

_____ 6. Plants that grow on the ocean floor belong to a group of organisms called benthos.

_____ 7. Squid and octopus live in the hadal ocean bottom zone.

_____ 8. The Delbuoy system of removing salt from ocean water developed by Pleass is powered by ocean waves.

_____ 9. In general, the elements needed by marine life are released at great depths in the ocean.

_____ 10. Pollutants can be found in measurable amounts everywhere in the oceans.

Choose the one best response. Write the letter of that choice in the space provided.

_____ 11. What percentage of ocean water is pure water?

      a. 3.5%   **b.** 39.5%   **c.** 57.5%   **d.** 96.5%

_____ 12. Which of the following gases dissolves most easily in ocean water?

      a. hydrogen   **b.** oxygen
      c. nitrogen   **d.** carbon dioxide

_____ 13. The most abundant major element dissolved in ocean water is:

      a. boron.   **b.** chlorine.   **c.** calcium.   **d.** potassium.

_____ 14. The most valuable resource taken from the ocean is:

      a. copper deposited on the ocean floor.
      b. calcite found in seashells.
      c. petroleum from beneath the ocean floor.
      d. gold dissolved in ocean water.

M O D E R N   E A R T H   S C I E N C E

Chapter 21

Choose the one best response. Write the letter of that choice in the space provided.

_____ 15. Which of the following elements is greatly depleted from ocean water by heavy plant growth?

    a. nitrogen   **b.** magnesium   **c.** oxygen   **d.** sodium

_____ 16. Nodules are a valuable source of:

    a. carbon.   **b.** manganese.   **c.** silicon.   **d.** magnesium.

_____ 17. In which of the following environments does salinity tend to be higher?

    a. surface temperate waters   **b.** surface tropical waters
    c. deep lake waters   **d.** deep tropical waters

_____ 18. Freezing ocean water can create fresh water because the salt:

    a. crystallizes within the ice.
    b. remains in pockets of liquid water.
    c. freezes faster than water.
    d. evaporates as the water is frozen.

Use the diagram below to answer questions 19 and 20.

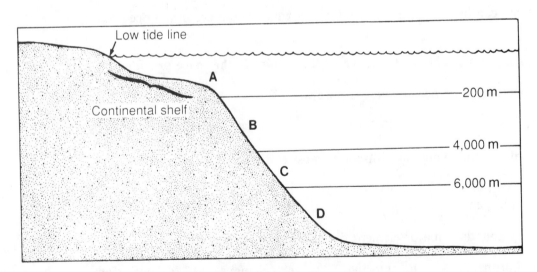

_____ 19. Which ocean environment in the diagram represents the sublittoral zone?

    a. A   **b. B**   **c. C**   **d. D**

_____ 20. Which ocean environment in the diagram represents the hadal zone?

    a. A   **b. B**   **c. C**   **d. D**

M O D E R N   E A R T H   S C I E N C E

Chapter 21

Read each question and answer it in the space provided.

21. What is defined by the number of grams of dissolved salt in 1 kg of ocean water?   _____

22. What color wavelengths of light tend to be reflected by ocean water?   _____

23. What is the process in which deep water moves upward to replace surface water?   _____

24. What are the free-floating, microscopic plants found in most regions of the near-surface ocean called?   _____

25. Which benthic zone is a relatively unstable environment for marine life but is populated by crabs, clams, and seaweed?   _____

26. What is the name for the larger swimming ocean animals that eat microscopic animals and plants?   _____

27. Which ocean environment extends seaward beyond the continental shelf?   _____

28. What is the general process of converting ocean water to fresh water called?   _____

29. What factor, besides the amount of dissolved solids, affects the density of water?   _____

30. What is the term for the farming of the ocean that involves developing and raising special breeds of marine plants and animals?   _____

Read each question or statement and answer it in the space provided.

31. Explain why a thermocline exists beneath much of the ocean surface.

_____

_____

_____

_____

_____

M O D E R N   E A R T H   S C I E N C E

Chapter 21

Read each question or statement and answer it in the space provided.

32. How has the addition of lead to gasoline affected marine life in the Pacific Ocean?

_____

_____

_____

_____

_____

33. How is the process of distillation used to remove salt from water?

_____

_____

_____

_____

_____

34. How does the process of evaporation affect the salinity of ocean water?

_____

_____

_____

_____

_____

35. How might a dramatic drop in sea level affect the current positions of the ocean environments?

_____

_____

_____

_____

_____

M O D E R N   E A R T H   S C I E N C E

Chapter 22
Movements of the Ocean

Read each statement below. If the statement is true, write _T_ in the space provided. If the statement is false, write _F_ in the space provided.

_____ **1.** Continental land masses act as barriers to surface currents.

_____ **2.** The Coriolis effect is caused by the earth's rotation.

_____ **3.** The overall pattern of the movement of surface currents in the Northern Hemisphere is clockwise.

_____ **4.** The Gulf Stream is a warm current.

_____ **5.** Longshore currents create a zigzag-shaped shoreline.

_____ **6.** Because of frictional pull of tides on the ocean floor, the earth's rotational speed is slowing.

_____ **7.** The tidal movement toward the coast is called the ebb tide.

_____ **8.** Tidal flats are largely unaffected by tidal currents.

_____ **9.** The winds in the Indian Ocean that change direction with each season are called typhoons.

_____ **10.** Deep currents tend to have cold, highly saline water in them.

Choose the one best response. Write the letter of that choice in the space provided.

_____ **11.** The global winds that are located just north and south of the equator are called the:

      a. rip currents.   **b.** westerlies.   **c.** trade winds.   **d.** monsoons.

_____ **12.** The direction of water particles in a wave in deep water can best be described as moving:

      a. forward.   **b.** in a long ellipse.
      c. up and down.   **d.** in a circle.

_____ **13.** The relatively weak, irregular current produced when the water from breaking waves is pulled back to deep water is called:

      a. drift.   **b.** a longshore current.
      c. an undertow.   **d.** a rip current.

Chapter 22

Choose the one best response. Write the letter of that choice in the space provided.

_____ **14.** Turbidity currents probably form as a result of:

      a. underwater landslides.   **b.** temperature differences.

      c. surface winds.   **d.** salinity differences.

_____ **15.** Which of the following occurs only during the full and new moon phases?

      a. spring tides   **b.** tidal oscillations

      c. neap tides   **d.** tidal bores

_____ **16.** Where are tidal currents generally strongest?

      a. along straight coastlines   **b.** in the open ocean

      c. along irregular coastlines   **d.** in a narrow bay

_____ **17.** Which of the following is the deepest current of the major oceans?

      a. Equatorial Countercurrent   **b.** Antarctic Bottom Water

      c. South Equatorial Current   **d.** North Atlantic Drift

_____ **18.** The speed at which a wave moves is calculated by:

      a. multiplying wavelength and period.

      b. dividing wavelength by period.

      c. multiplying wave period and height.

      d. dividing wave period by height.

Use the diagram below to answer questions 19 and 20.

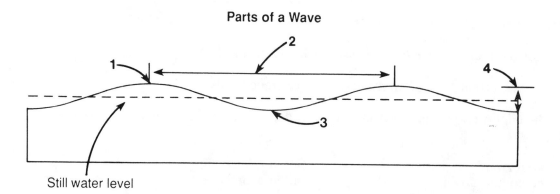

Parts of a Wave

Still water level

_____ **19.** Which number in the diagram indicates a wave crest?

    a. 1   **b.** 2   **c.** 3   **d.** 4

_____ **20.** Which number in the diagram indicates a wavelength?

    a. 1   **b.** 2   **c.** 3   **d.** 4

M O D E R N   E A R T H   S C I E N C E

Chapter 22

Read each question and answer it in the space provided.

21. What general term is used to describe the type of current that is driven by the wind? _____

22. In what body of water does the tidal range exceed 15 m due to tidal oscillations. _____

23. What is the term for the time it takes for one complete wavelength to pass a fixed point? _____

24. What is the bending of waves called? _____

25. What is another name for a seismic sea wave? _____

26. What are the daily changes in the level of an ocean surface called? _____

27. Which planetary body has the most influence on the earth's ocean tides? _____

28. What is formed when the crest is blown off of a wave? _____

29. What is the term for the difference between high tide and low tide at a given location? _____

30. What is the term used to describe the distance that wind has traveled across open water? _____

Read each question or statement and answer it in the space provided.

31. What causes a wave to break?

_____

_____

_____

_____

_____

M O D E R N   E A R T H   S C I E N C E

Chapter 22

Read each question or statement and answer it in the space provided.

32. Explain the reason for the difference in density of warm and cold water.

_____

_____

_____

_____

_____

33. Describe how the wave height of a tsunami changes while traveling from open ocean to the shore.

_____

_____

_____

_____

_____

34. What causes a tide on the side of earth opposite the moon to occur?

_____

_____

_____

_____

_____

35. Would it probably be better to plan the site of an oceanside building during a period of spring or neap tides? Explain your answer.

_____

_____

_____

_____

_____

M O D E R N   E A R T H   S C I E N C E

Unit 6
Oceans

Read each statement below. If the statement is true, write *T* **in the space provided.**
If the statement is false, write *F* **in the space provided.**

_____ 1. Guyots are formed by the erosion and subsidence of seamounts.

_____ 2. The oceans contain about three fourths of the earth's water.

_____ 3. Samples of deep-sea sediments are often obtained by taking core samples.

_____ 4. Some nations that formerly had abundant supplies of fresh water are now facing shortages.

_____ 5. The physical properties of ocean water are those characteristics that determine the water's composition and enable the water to dissolve other substances.

_____ 6. The deep currents of the Atlantic Ocean tend to flow in the same direction as the Atlantic surface currents.

_____ 7. The energy of tidal movements has been harnessed to generate electricity.

_____ 8. The salinity of ocean water tends to be higher in deep-ocean waters.

_____ 9. The last color of visible light to be absorbed by ocean water is blue.

_____ 10. Abyssal plains are the flattest regions on earth.

_____ 11. The effects of tidal oscillations are most apparent along straight coastlines.

_____ 12. Plant growth is present in all ocean environments.

_____ 13. The tide occurs on the sides of the earth that face toward and away from the moon.

_____ 14. Wind produces both surface currents and waves.

_____ 15. The ocean floor is composed mainly of continental crust.

_____ 16. The finest ocean sediments are found closest to shore.

M O D E R N   E A R T H   S C I E N C E

Unit 6

Choose the one best response. Write the letter of that choice in the space provided.

_____ **17.** Which of the following is one of a group of long, rolling waves that are all the same size?

        a. drift   **b.** fetch   **c.** swell   **d.** crest

_____ **18.** Fracture zones are most often found near:

        a. trenches.   **b.** the continental shelf.
        c. mid-ocean ridges.   **d.** the continental slope.

_____ **19.** Which of the following is a part of three other oceans?

        a. Antarctic Ocean   **b.** Arctic Ocean
        c. Indian Ocean   **d.** Atlantic Ocean

_____ **20.** Turbidity currents often result in the formation of a:

        a. submarine canyon.   **b.** guyot.
        c. continental slope.   **d.** fracture zone.

_____ **21.** Which of the following is one of the three principal gases dissolved in ocean water?

        a. hydrogen   **b.** nitrogen   **c.** carbon   **d.** sodium

_____ **22.** When a river enters the ocean through a long bay, the ocean tide may rush into the river, creating a surge of water called a tidal:

        a. bore.   **b.** oscillation.   **c.** range.   **d.** flat.

_____ **23.** Whitecaps are caused by:

        a. wind during storms.   **b.** longshore currents.
        c. refraction of waves.   **d.** tidal currents.

_____ **24.** Which of the following sediments would most likely be found in a sample from the ocean floor at a depth of 5,000 m?

        a. volcanic dust   **b.** siliceous ooze
        c. rock fragments   **d.** calcareous ooze

_____ **25.** Which of these factors is used to calculate wave speed?

        a. wind speed   **b.** wave period   **c.** water depth   **d.** wave height

_____ **26.** Tides are primarily due to:

        a. density differences.   **b.** currents.
        c. salinity differences.   **d.** gravitational attraction.

M O D E R N   E A R T H   S C I E N C E

Unit 6

Choose the one best response. Write the letter of that choice in the space provided.

_____ **27.** Nodules found on the ocean floor appear to be the result of:

    a. volcanic activity.   **b.** chemical reactions.
    c. turbidity currents.   **d.** transform faulting.

Use the diagram below to answer question 28.

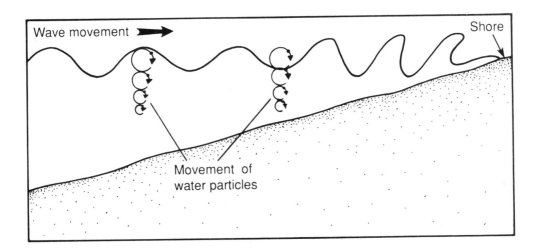

_____ **28.** Which of the following occurs when waves approach a shoreline as shown in this diagram?

    a. wavelength increases   **b.** wave period increases
    c. wave height decreases   **d.** wave height increases

_____ **29.** Organisms that live on the ocean floor are called:

    a. zooplankton.   **b.** nekton.
    c. benthos.   **d.** dolphins.

_____ **30.** Organisms that live in the intertidal zone include:

    a. clams and crabs.   **b.** squid and octopus.
    c. whales and devilfish.   **d.** sea cucumbers and tube worms.

_____ **31.** Which of the following is extracted easily from the ocean and refined to produce metal?

    a. magnesium   **b.** phosphate   **c.** gold   **d.** sulfur

MODERN EARTH SCIENCE

Unit 6

Read each question and answer it in the space provided.

32. Which pelagic zone is the source of much of the fish and seafood that people eat?

      _____

33. What global winds are located just north and south of the equator?

      _____

34. What is the name for a foamy mass of water created when waves topple over?

      _____

35. What type of submersible was used to photograph the inside of the remains of the *Titanic*?

      _____

36. What type of ocean-floor sediment is formed by the remains of radiolaria?

      _____

37. Which of the earth's principal oceans is deepest?

      _____

38. What is the group of microscopic organisms called that use the sun's light to produce food through photosynthesis?

      _____

39. What type of tide occurs when the sun, moon, and earth are all aligned?

      _____

40. What short-lived current is produced by underwater landslides?

      _____

41. What are the skeletons of foraminifera and corals composed of?

      _____

42. Which ocean environment is characterized by areas that are rich in minerals and an ocean floor marked by volcanic vents?

      _____

43. What is the method of desalination called that involves heating ocean water to remove its salt?

      _____

44. What major physical property of ocean water is affected by water temperature and the amount of dissolved solids in the water?

      _____

M O D E R N E A R T H S C I E N C E

Unit 6

Read each question or statement and answer it in the space provided.

45. How did the continental shelves become flattened?

_____

_____

_____

_____

_____

_____

46. Describe the movement of water particles in a wave during one wave period.

_____

_____

_____

_____

_____

_____

47. Explain how fresh water can be extracted from ocean water by freezing the ocean water.

_____

_____

_____

_____

_____

_____

M O D E R N   E A R T H   S C I E N C E

Unit 6

Read each question or statement and answer it in the space provided.

48. Describe a thermocline in terms of where and why it occurs.

_____

_____

_____

_____

_____

_____

49. How do the tidal patterns on the Atlantic Coast and the Pacific Coast of the United States differ?

_____

_____

_____

_____

_____

_____

50. How might meteorite fragments become part of the ocean-floor sediments?

_____

_____

_____

_____

_____

_____

M O D E R N   E A R T H   S C I E N C E

Chapter 23
The Atmosphere

Read each statement below. If the statement is true, write _T_ in the space provided. If the statement is false, write _F_ in the space provided.

_____ **1.** Valley breezes blow from mountains into valleys.

_____ **2.** Standard atmospheric pressure is equal to a reading of 760 mm of mercury on a barometer.

_____ **3.** The process by which the earth's atmosphere traps infrared rays is called the greenhouse effect.

_____ **4.** Auto emissions are a major contributor to air pollution.

_____ **5.** Most of the ozone in the earth's atmosphere is located in the ionosphere.

_____ **6.** Atmospheric pressure is generally higher under a body of warm air than under a body of cool air.

_____ **7.** Scattering is responsible for making the sky appear blue.

_____ **8.** Air pressure increases with elevation.

_____ **9.** The Darrieus turbine generates power from the wind.

_____ **10.** An aneroid barometer can be used as an altimeter.

Choose the one best response. Write the letter of that choice in the space provided.

_____ **11.** Ozone is a form of:

    a. nitrogen.   **b.** carbon dioxide.
    c. oxygen.   **d.** water vapor.

_____ **12.** The vertical movement of air due to uneven heating is called:

    a. convection.   **b.** refraction.   **c.** conduction.   **d.** radiation.

_____ **13.** The layer of the atmosphere in which weather change occurs is the:

    a. stratopause.   **b.** troposphere.
    c. mesopause.   **d.** thermosphere.

_____ **14.** The coldest layer of the atmosphere is the:

    a. mesosphere.   **b.** exosphere.   **c.** stratosphere.   **d.** ionosphere.

_____ **15.** The albedo of the earth's surface is the fraction of solar radiation that is:

    a. scattered.   **b.** reflected.   **c.** inverted.   **d.** absorbed.

M O D E R N   E A R T H   S C I E N C E

Chapter 23

Choose the one best response. Write the letter of that choice in the space provided.

_____ **16.** The most abundant element in the earth's atmosphere is:

        a. oxygen.   **b.** argon.   **c.** hydrogen.   **d.** nitrogen.

_____ **17.** Which of the following adds oxygen to the atmosphere?

        a. forest fires   **b.** weathering of rocks
        c. photosynthesis   **d.** life processes of animals

Use the diagram below to answer questions 18-20.

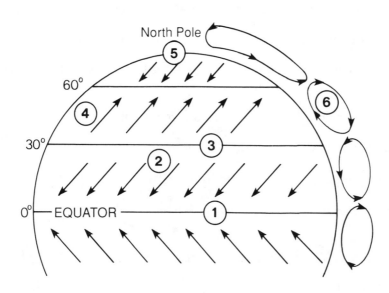

_____ **18.** The jet streams usually form as a result of interaction between zones **4** and:

        a. 6.   **b.** 5.   **c.** 3.   **d.** 2.

_____ **19.** Which winds are represented by the arrows labeled **4**?

        a. southeast trades   **b.** westerlies
        c. northeast trades   **d.** polar easterlies

_____ **20.** Which number represents a convection cell?

        a. 1   **b.** 3   **c.** 5   **d.** 6

M O D E R N   E A R T H   S C I E N C E

Chapter 23

Complete each statement by writing the correct term or phrase in the space provided.

21. Acid precipitation is produced primarily by the burning of _____ .

22. Gentle winds that extend over distances of less than 100 km are called

   _____ .

23. The deflection of winds by the earth's rotation is called the _____ .

24. The outermost layer of the earth's atmosphere is the _____ .

25. Objects on earth are heated when they absorb the sun's visible light rays and its

   _____ .

26. The general weather conditions over many years is called

   _____ .

27. The condition in which cold air is trapped under warm air is called

   _____ .

28. The bands of high-speed winds found in the upper troposhere and lower stratosphere are

   called _____ .

29. The units used on weather maps to indicate atmospheric pressure are called

   _____ .

30. The form of heat transfer in which fast-moving molecules cause other molecules to also

   move faster is called _____ .

Read each question or statement and answer it in the space provided.

31. Why is the hottest part of the day usually at midafternoon rather than at noon when the sun
   is highest in the sky?

   _____

   _____

   _____

   _____

   _____

M O D E R N   E A R T H   S C I E N C E

Chapter 23

Read each question or statement and answer it in the space provided.

32. Why is the apparent deterioration of the ozone layer a major concern?

_____

_____

_____

_____

_____

33. What are two harmful effects of air pollution on humans?

_____

_____

_____

_____

_____

34. Explain how a sea breeze is replaced by a land breeze at night.

_____

_____

_____

_____

_____

35. From where does the light of the full moon originate, and by what process is it directed toward the earth?

_____

_____

_____

_____

_____

M O D E R N   E A R T H   S C I E N C E

Chapter 24
Water in the Atmosphere

Read each statement below. If the statement is true, write *T* **in the space provided.**
If the statement is false, write *F* **in the space provided.**

_____ **1.** Most evaporation takes place in regions around the equator.

_____ **2.** Cold air can hold more water vapor than warm air.

_____ **3.** The dew point temperature depends on the amount of water in the air.

_____ **4.** Air must be saturated before clouds can form.

_____ **5.** The lowest clouds in the sky are cirrus clouds.

_____ **6.** Fog does not require condensation nuclei in order to form.

_____ **7.** Hail forms when rain falls through a layer of freezing air.

_____ **8.** In tropical regions, rain is commonly the result of the process of coalescence.

_____ **9.** Silver-iodide vapor and powdered dry ice have been used to cause or increase precipitation artificially.

_____ **10.** Rain gauges measure the precipitation that falls over a large region.

Choose the one best response. Write the letter of that choice in the space provided.

_____ **11.** Ice changes directly into water vapor by the process of:

    a. evaporation.   **b.** sublimation.
    c. condensation.   **d.** saturation.

_____ **12.** What is the relative humidity when there are 7 g/m³ of water vapor in air with a saturation point of 14 g/m³?

    a. 7%   **b.** 14%   **c.** 50%   **d.** 98%

_____ **13.** What forms when the dew point is below 0°C?

    a. dew   **b.** fog   **c.** frost   **d.** drizzle

_____ **14.** As air rises and expands, it undergoes:

    a. advective heating.   **b.** advective cooling.
    c. convective heating.   **d.** convective cooling.

Choose the one best response. Write the letter of that choice in the space provided.

_____ **15.** Which type of fog usually forms over inland rivers and lakes?

      a. radiation   **b.** steam   **c.** upslope   **d.** advection

_____ **16.** The most common form of solid precipitation is:

      a. glaze ice.   **b.** hail.   **c.** sleet.   **d.** snow.

_____ **17.** Most of the water in supercooled clouds exists as:

      a. snowflakes.   **b.** water droplets.
      c. ice crystals.   **d.** water vapor.

_____ **18.** A funnel and a cylindrical container could be used to measure the:

      a. amount of rainfall.   **b.** specific humidity.
      c. adiabatic temperature change.   **d.** dew point.

Use the diagram below to answer questions 19 and 20.

_____ **19.** What type of cloud is pictured in the diagram?

      a. cumulus   **b.** stratus   **c.** cirrus   **d.** nimbus

_____ **20.** At what altitude would you expect to find this type of cloud?

      a. 1,000 m   **b.** 2,000 m   **c.** 4,000 m   **d.** 6,000 m

Complete each statement by writing the correct term or phrase in the space provided.

21. The heat released when water condenses and changes to liquid form is called

_____ .

22. The actual amount of water vapor in a volume of air is expressed as specific

_____ .

M O D E R N   E A R T H   S C I E N C E

Chapter 24

Complete each statement by writing the correct term or phrase in the space provided.

23. When air contacts a cold surface, the air may cool to its dew point by the process of

_____ .

24. Water vapor condenses on suspended particles called _____ .

25. Clouds produced by the rising and cooling of large bodies of air are called

_____ .

26. Radiation fog is also called _____ .

27. Snowflakes vary in size depending on the air _____ .

28. The type of condensation nucleus that has a crystalline structure similar to ice is called

_____ .

29. Droughts may eventually be ended using a rain-producing method called

_____ .

30. Snow is measured by both the depth of accumulation and the _____ .

Read each statement and answer it in the space provided.

31. Describe the differences in molecular motion found in ice, water, and water vapor.

_____

_____

_____

_____

_____

32. Describe how a psychrometer works.

_____

_____

_____

_____

_____

M O D E R N   E A R T H   S C I E N C E

Chapter 24

Read each statement and answer it in the space provided.

33. Explain how precipitation forms in supercooled clouds.

_____

_____

_____

_____

_____

34. Describe what the inside of a hailstone looks like and what has caused this appearance.

_____

_____

_____

_____

_____

35. Explain why drops of water may form on the outside of a glass filled with ice water.

_____

_____

_____

_____

_____

M O D E R N   E A R T H   S C I E N C E

Chapter 25
Weather

Read each statement below. If the statement is true, write _T_ in the space provided. If the statement is false, write _F_ in the space provided.

_____ 1. A large body of air with highly variable temperature and moisture content is called an air mass.

_____ 2. Three polar air masses influence weather in North America.

_____ 3. A squall line is a long line of heavy thunderstorms.

_____ 4. In North America, wave cyclones generally move in a westerly direction and spin clockwise.

_____ 5. Hurricanes are more powerful than wave cyclones.

_____ 6. An anemometer measures wind direction.

_____ 7. Radar is often used to detect cloud droplets and dust particles.

_____ 8. Lines on a weather map connecting points of equal humidity are called isobars.

_____ 9. To forecast the weather, meteorologists compare the most recent weather map with maps from the previous 24 hours.

_____ 10. Meteorological technicians must have advanced college degrees in meteorology.

Choose the one best response. Write the letter of that choice in the space provided.

_____ 11. What letters would be used to designate an air mass that forms over Antarctica?

     a. cP   **b.** cT   **c.** mP   **d.** mT

_____ 12. During which season do continental tropical air masses flow over North America?

     a. autumn   **b.** spring   **c.** summer   **d.** winter

_____ 13. What type of front forms when a warm air mass overtakes a cold air mass?

     a. cold   **b.** warm   **c.** stationary   **d.** occluded

_____ 14. Lightning causes a rapid expansion and collapse of the air that produces:

     a. thunder.   **b.** hail.   **c.** cyclones.   **d.** rain.

M O D E R N   E A R T H   S C I E N C E

Chapter 25

Choose the one best response. Write the letter of that choice in the space provided.

_____ **15.** The process of adding freezing nuclei to supercooled clouds is called:

    a. ionizing.   **b.** seeding.   **c.** jetting.   **d.** linking.

_____ **16.** Which of the following instruments is used to measure temperature changes?

    a. barometer   **b.** hygrometer   **c.** psychrometer   **d.** thermograph

_____ **17.** On a weather map, clusters of symbols showing weather conditions around an observation center are called:

    a. station models.   **b.** atmospheric clusters.
    c. observation systems.   **d.** forecasting points.

_____ **18.** The center of an anticyclone is an area of:

    a. cold temperatures.   **b.** warm temperatures.
    c. low pressure.   **d.** high pressure.

Use the diagram below to answer questions 19 and 20.

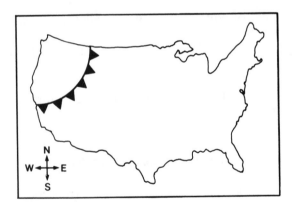

_____ **19.** What feature is shown on this weather map?

    a. warm front   **b.** hurricane   **c.** cold front   **d.** typhoon

_____ **20.** The feature on the map is moving toward the:

    a. northwest.   **b.** northeast.   **c.** southwest.   **d.** southeast.

Read each question and answer it in the space provided.

21. What type of air mass forms over the northern Pacific and southwestern Alaska? _____

22. Which type of front forms when two air masses meet, but neither of them is displaced? _____

M O D E R N   E A R T H   S C I E N C E

Chapter 25

Read each question and answer it in the space provided.

23. In which direction do wave cyclones spin in the Northern Hemisphere?

_____

24. How is a south wind indicated in degrees?

_____

25. What is the general term for air masses that form over the ocean?

_____

26. Local daily extended forecasts usually predict weather conditions for how many days in advance?

_____

27. The release of large quantities of ions near the ground can modify the electrical properties of small cumulus clouds to help control what?

_____

28. What type of meteorological instrument package is carried into the atmosphere by helium-filled balloons to investigate weather conditions?

_____

29. Where are cameras mounted to take images of cloud cover over all of North America?

_____

30. What type of front forms when cold air lifts warm air, causing the warm air to be completely cut off from the ground?

_____

Read each question or statement and answer it in the space provided.

31. How do tornadoes develop?

_____

_____

_____

_____

32. Explain how convection cells form in the Northern and Southern Hemispheres.

_____

_____

_____

_____

M O D E R N   E A R T H   S C I E N C E

Chapter 25

Read each question or statement and answer it in the space provided.

33. What are the three stages of a thunderstorm? How do these storms form?

_____

_____

_____

_____

34. Describe how a liquid thermometer works.

_____

_____

_____

_____

35. What techniques do scientists use to control lightning?

_____

_____

_____

_____

M O D E R N   E A R T H   S C I E N C E

Chapter 26
Climate

Read each statement below. If the statement is true, write *T* in the space provided. If the statement is false, write *F* in the space provided.

_____ **1.** The bora is a warm wind that comes down from the mountains.

_____ **2.** On lakes, waves and currents prevent surface temperatures from changing rapidly.

_____ **3.** Tropical climates are located near the equator.

_____ **4.** Subarctic climates have small yearly temperature ranges.

_____ **5.** Most of the rainfall in middle-latitude areas results from wave cyclones.

_____ **6.** Mass destruction of the earth's rain forests may be causing a rise in global temperatures.

_____ **7.** Lakes may have a moderating effect on local climates.

_____ **8.** Forests affect local climates by increasing the humidity.

_____ **9.** El Niño causes a cooling of the waters off the west coast of South America.

_____ **10.** Water releases heat faster than does land.

Choose the one best response. Write the letter of that choice in the space provided.

_____ **11.** A region that receives 8 hours of sunlight in December and 16 hours of light in June is located at about:

    a. the equator.
    c. the North Pole.
    b. 45° north latitude.
    d. 45° south latitude.

_____ **12.** Specific heat is the amount of heat needed to raise the temperature of 1 g of a substance by:

    a. 1°C.   **b.** 5°C.   **c.** 10°C.   **d.** 50°C.

_____ **13.** Which of the following blows down from the Alps to the Mediterranean Sea?

    a. chinook   **b.** mistral   **c.** foehn   **d.** bora

_____ **14.** In which climate would dense forests of cone-bearing trees most likely grow?

    a. tundra
    c. tropical savanna
    b. marine west-coast
    d. middle-latitude desert

Chapter 26

Choose the one best response. Write the letter of that choice in the space provided.

_____ **15.** Areas poleward of subarctic climates are characterized by their:

    a. large daily temperature ranges.
    b. lack of vegetation.
    c. small amounts of precipitation.
    d. vast areas of sand.

_____ **16.** Cities usually receive more rainfall than nearby rural areas as a result of having:

    a. less radiational cooling at night.
    b. less vegetation to absorb solar radiation.
    c. increased energy usage in cities.
    d. increased numbers of condensation nuclei.

_____ **17.** The climate in which the vegetation consists of lichen, mosses, and small flowering plants is called:

    a. tundra   **b.** middle-latitude steppe
    c. tropical savanna   **d.** tropical desert

_____ **18.** Which climate is characterized by mild winters and warm, humid summers?

    a. humid continental   **b.** humid subtropical
    c. Mediterranean   **d.** tropical rain forest

Use the map below to answer questions 19 and 20.

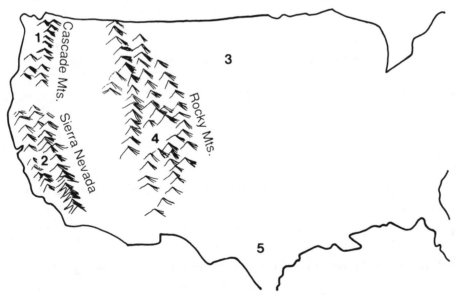

_____ **19.** Which region in the map has a middle-latitude steppe climate?

    a. 1   **b.** 2   **c.** 3   **d.** 5

_____ **20.** Which region has the greatest average yearly rainfall?

    a. 1   **b.** 2   **c.** 3   **d.** 4

M O D E R N   E A R T H   S C I E N C E

Chapter 26

Complete each statement by writing the correct term or phrase in the space provided.

21. Between the tropical and polar climate zones is the zone of temperate climates called

_____ .

22. The most important factor affecting local weather conditions is the area's

_____ .

23. The climates typically found above 55° north latitude are called _____ .

24. The wind with a seasonal shift in direction common in equatorial regions having long

coastlines is called _____ .

25. When a moving air mass encounters a mountain range, it rises, expands, and

_____ .

26. The sun's rays strike the earth's surface at almost a 90° angle near the

_____ .

27. The two weather factors most often used to describe climate are precipitation and average

_____ .

28. Each principal climate zone is divided into several different types of climates because of

differences in the amount of _____ .

29. The equatorial belt of low pressure and heavy rainfall is called the

_____ .

30. The region roughly located between the Tropic of Cancer and the Tropic of Capricorn is the

type of climate zone called _____ .

Read each question or statement and answer it in the space provided.

31. On one day, a city has a high temperature of 30°C and a low temperature of 12°C. What are
the average daily temperature and temperature range of the city?

_____

_____

_____

_____

M O D E R N   E A R T H   S C I E N C E

Chapter 26

Read each question or statement and answer it in the space provided.

32. Explain how the wind patterns of southern and eastern Asia change with the seasons.

_____

_____

_____

_____

33. Why does the Gulf Stream have a moderating effect on the climate of northwestern Europe but little effect on the climate of the east coast of the United States?

_____

_____

_____

_____

34. What causes the tropical savanna climate to vary at different times of year?

_____

_____

_____

_____

35. Death Valley lies east of the Sierra Nevada of California and has a tropical desert climate, while about 300 km to the west, in Los Angeles, the climate is Mediterranean. Explain the difference in climate.

_____

_____

_____

_____

M O D E R N   E A R T H   S C I E N C E

Unit 7
Atmospheric Forces

Read each statement below. If the statement is true, write *T* in the space provided. If the statement is false, write *F* in the space provided.

_____ 1. Ocean currents have a stronger effect on land air masses when winds consistently blow away from the shore.

_____ 2. Mosses, lichens, and small flowering plants are common in areas of tundra climate.

_____ 3. Sublimation occurs only when the temperature is above freezing.

_____ 4. Condensing water vapor slows the rate of adiabatic temperature change.

_____ 5. Most air pollution comes from the burning of coal and petroleum fuels.

_____ 6. During daylight hours, there is little temperature difference between a body of water and the land along its shore.

_____ 7. Fronts form when two air masses with identical characteristics merge.

_____ 8. Wind direction may be described by one of 32 points on the compass.

_____ 9. Differences in air pressure and temperature cause convection cells to form.

_____ 10. Cities are generally cooler than nearby rural areas.

_____ 11. The amount of solar radiation received by a region is determined primarily by the latitude of the region.

_____ 12. Solid materials are good conductors of heat.

_____ 13. A mirage is created by the refraction of light rays.

_____ 14. Electric thermometers can be used to measure temperature changes when an observer is not present.

_____ 15. Glaze ice forms when rain freezes as it strikes the ground.

Choose the one best response. Write the letter of that choice in the space provided.

_____ 16. Which of the following changes as air temperature changes?

     a. dew point   **b.** relative humidity
     c. freezing point   **d.** specific humidity

_____ 17. What type of cloud is composed entirely of ice crystals?

     a. stratus   **b.** nimbus   **c.** cirrus   **d.** cumulus

Unit 7

Choose the one best response. Write the letter of that choice in the space provided.

_____ **18.** Which of the following is a warm, dry wind that flows down the Alps?

      a. mistral   **b.** bora   **c.** foehn   **d.** monsoon

_____ **19.** Which of these climates has wet summers and dry winters?

      a. middle-latitude steppe   **b.** tropical savanna
      c. Mediterranean   **d.** tundra

_____ **20.** Which of the following phenomena is responsible for red sunsets?

      a. reflection   **b.** albedo   **c.** convection   **d.** scattering

_____ **21.** What type of radiation is trapped on the earth's surface by the greenhouse effect?

      a. ultraviolet rays   **b.** X rays
      c. gamma rays   **d.** infrared rays

Use the diagram below to answer question 22.

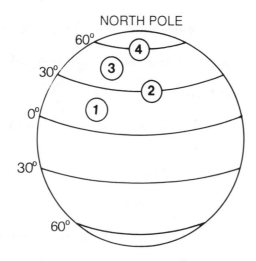

_____ **22.** Which number represents the horse latitudes?

      a. 1   **b.** 2   **c.** 3   **d.** 4

_____ **23.** Which type of air mass causes heavy precipitation in Oregon and Washington State?

      a. maritime polar Atlantic   **b.** continental tropical
      c. maritime polar Pacific   **d.** maritime tropical gulf

_____ **24.** Large, low-pressure storm centers that form along polar fronts and influence weather in the middle latitudes are called:

      a. Coriolis formations.   **b.** jet-stream systems.
      c. wave cyclones.   **d.** polar storm fronts.

M O D E R N   E A R T H   S C I E N C E

Unit 7

Choose the one best response. Write the letter of that choice in the space provided.

_____ **25.** Rain in tropical regions is commonly formed by the process of:

   a. supercooling.   **b.** convection.   **c.** sublimation.   **d.** coalescence.

_____ **26.** Which of the following properties of snow is commonly determined by melting a sample of snow and measuring the result?

   a. water content   **b.** depth   **c.** crystal size   **d.** density

Use the map below to answer questions 27 and 28.

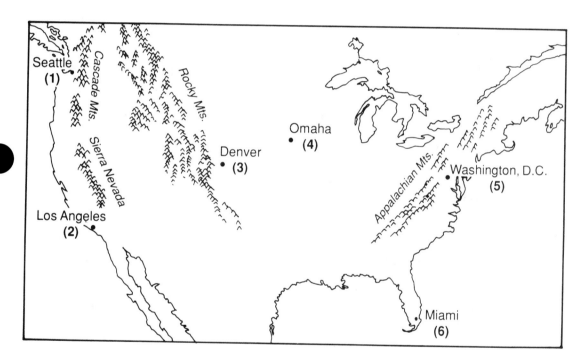

_____ **27.** Which of these cities has a humid continental climate?

   a. 1   **b.** 3   **c.** 5   **d.** 6

_____ **28.** Which of these cities has a humid subtropical climate?

   a. 2   **b.** 4   **c.** 5   **d.** 6

M O D E R N   E A R T H   S C I E N C E

Unit 7

Choose the one best response. Write the letter of that choice in the space provided.

_____ **29.** What would be the long-range effect on the earth's surface if the ozone layer were destroyed?

    a. decreased infrared radiation
    b. increased acid precipitation
    c. increased ultraviolet radiation
    d. decreased atmospheric temperature

Use the diagram below to answer question 30.

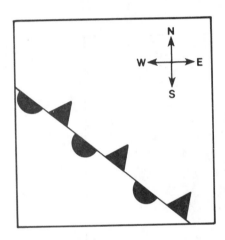

_____ **30.** What type of front is shown in this diagram?

    a. stationary   **b.** cold
    c. warm   **d.** occluded

Read each question and answer it in the space provided.

31. What letter designation is used for air masses that form over northern Canada? _____

32. In which direction do wave cyclones generally move in North America? _____

33. What type of fog usually forms on calm, clear nights and results from the nightly cooling of the earth? _____

34. What is the term for the amount of heat needed to raise the temperature of 1 g of a substance by 1°C? _____

35. A violent storm with a funnel-shaped cloud and winds of over 400 km/hr occurs over the ocean. What is this type of storm called? _____

36. What layer of the atmosphere can reflect radio waves back to the earth? _____

M O D E R N   E A R T H   S C I E N C E

Unit 7

Read each question and answer it in the space provided.

37. What is the term for the temperature to which air must be cooled to reach saturation?   _____

38. What general weather condition is most difficult to predict accurately?   _____

39. What instrument measures atmospheric pressure?   _____

40. What is the name for the mixture of gases and particles that surrounds the earth?   _____

41. What is the name given to the densely vegetated tropical areas located close to the equator?   _____

42. What instrument package is carried in a balloon to high altitudes in order to measure temperature and humidity?   _____

43. Which polar climate zone is characterized by a lack of trees?   _____

44. What term is used to express the actual amount of moisture in air?   _____

Read each question or statement and answer it in the space provided.

45. Describe the seasonal variations in the location of a polar front.

_____

_____

_____

_____

_____

46. Describe the Chinooks and where they are located.

_____

_____

_____

_____

_____

M O D E R N   E A R T H   S C I E N C E

Unit 7

Read each question or statement and answer it in the space provided.

47. How does the process of cloud seeding work?

_____

_____

_____

_____

_____

48. Describe how the process of convection helps warm the earth's surface.

_____

_____

_____

_____

_____

49. Explain how hailstones form.

_____

_____

_____

_____

_____

50. What causes a sunset to be red?

_____

_____

_____

_____

_____

M O D E R N   E A R T H   S C I E N C E

Chapter 27
Stars and Galaxies

Read each statement below. If the statement is true, write _T_ in the space provided. If the statement is false, write _F_ in the space provided.

_____ 1. A main-sequence star maintains a stable size as long as there is enough hydrogen to fuse into helium.

_____ 2. A cloud of dust and gas from which a star forms is called a nebula.

_____ 3. All stars have the same composition.

_____ 4. The Milky Way Galaxy does not rotate.

_____ 5. Quasars have large red shifts.

_____ 6. A globular cluster is a type of galaxy.

_____ 7. Most constellations look like the figures for which they were named.

_____ 8. A star with a surface temperature of 25,000°C would be a red star.

_____ 9. Elliptical galaxies contain very little dust and gas.

_____ 10. The Milky Way Galaxy is an example of an irregular galaxy.

Choose the one best response. Write the letter of that choice in the space provided.

_____ 11. An apparent shift in the wavelength of light emitted by a light source moving toward or away from an observer is known as:

    a. spectrum analysis.   **b.** parallax.
    c. the Doppler effect.   **d.** a Cepheid variable.

_____ 12. A pulsar is a type of:

    a. protostar.   **b.** neutron star.   **c.** white dwarf.   **d.** supergiant.

_____ 13. Astronomers label the stars within constellations according to:

    a. composition.   **b.** color.
    c. surface temperature.   **d.** apparent magnitude.

_____ 14. A star that has a blue-shifted spectrum is most likely moving:

    a. toward the earth.   **b.** away from the sun.
    c. around the Milky Way Galaxy.   **d.** toward a black hole.

M O D E R N   E A R T H   S C I E N C E

Chapter 27

Choose the one best response. Write the letter of that choice in the space provided.

_____ **15.** A star with which of the following apparent magnitudes would appear brightest?

    a. 10   **b.** 5   **c.** 1   **d.** –5

_____ **16.** Which of the following stages in a star's evolution follows fusion of heavy elements into iron?

    a. white dwarf   **b.** supernova
    c. black hole   **d.** main-sequence

_____ **17.** Large-scale groups of stars bound together by gravitational attraction are known as:

    a. galaxies.   **b.** multiple-star systems.
    c. Cepheids.   **d.** clusters.

_____ **18.** Which of the following stages is the earliest in the development of a star?

    a. neutron star   **b.** protostar   **c.** dark nebula   **d.** giant

Use the diagram below to answer questions 19 and 20.

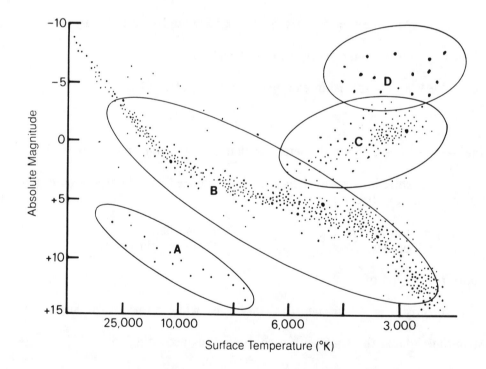

_____ **19.** What type of star is generally located in region **A** of this Hertzsprung-Russell diagram?

    a. main-sequence   **b.** protostar
    c. giant   **d.** white dwarf

_____ **20.** In which region of the diagram does the sun lie?

    a. A   **b.** B   **c.** C   **d.** D

M O D E R N   E A R T H   S C I E N C E

Chapter 27

Complete each statement by writing the correct term or phrase in the space provided.

21. Patterns of stars that appear relatively fixed in the night sky are known as

_____ .

22. The most distant objects that have been observed from the earth are called

_____ .

23. The process through which smaller atomic nuclei are combined into larger ones is known as

_____ .

24. Spherical star clusters that are distributed around the central core of the galaxy are called

_____ .

25. The only direct method scientists use to measure the distance from the earth to stars is known

as _____ .

26. The Milky Way Galaxy and about seventeen other galaxies within three million light-years of

the Milky Way Galaxy are collectively known as the _____ .

27. The phase of stellar evolution that is characterized by fusion of hydrogen atoms into helium

atoms is called the _____ .

28. The true brightness of a star is known as its _____ .

29. A star that is always visible in the night sky is called _____ .

30. The distance between a star and the earth is measured in _____ .

Read each question or statement and answer it in the space provided.

31. How do astronomers try to locate black holes?

_____

_____

_____

_____

MODERN EARTH SCIENCE

Chapter 27

Read each question or statement and answer it in the space provided.

32. A photograph of the motion of stars in the northern sky during an extended period shows a pattern of curved trails. Explain what causes this apparent motion of the stars.

_____

_____

_____

_____

33. Explain the big bang theory of the formation of the universe.

_____

_____

_____

_____

34. Explain how scientists use a Cepheid variable star to measure the distance to the star's galaxy.

_____

_____

_____

_____

35. The core of a star is five times more massive than the sun and composed mostly of helium and carbon. Which stage of stellar evolution is the star most likely in? Why?

_____

_____

_____

_____

M O D E R N   E A R T H   S C I E N C E

Chapter 28
The Sun

Read each statement below. If the statement is true, write *T* in the space provided. If the statement is false, write *F* in the space provided.

_____ **1.** Auroras usually occur near sunspots.

_____ **2.** Prominences follow curved lines of magnetic force.

_____ **3.** During the hydrogen fusion reaction, the mass of the two original hydrogen nuclei is more than the mass of the fused nucleus.

_____ **4.** The nebular theory states that the sun formed before the planets.

_____ **5.** Planetesimals coalesced into the protoplanets.

_____ **6.** The sun's energy output remains constant over time.

_____ **7.** Water absorbs carbon dioxide from the atmosphere.

_____ **8.** The radiative zone is the hottest region of the sun.

_____ **9.** The nebular theory was developed from Laplace's hypothesis.

_____ **10.** During the formation of the solar system, the distance between the developing sun and protoplanets influenced the composition of the planets that formed.

Choose the one best response. Write the letter of that choice in the space provided.

_____ **11.** The sunspot cycle lasts an average of:

      a. 2 years.   **b.** 11 years.   **c.** 20 years.   **d.** 37 years.

_____ **12.** One of the most violent of all solar disturbances is called a:

      a. solar flare.   **b.** prominence.   **c.** solar wind.   **d.** sunspot.

_____ **13.** Which of the following planets is a gas giant?

      a. Uranus   **b.** Venus   **c.** Mars   **d.** Mercury

_____ **14.** Einstein's equation $E = mc^2$ can be used to calculate the amount of energy produced from a given amount of:

      a. heat.   **b.** volume.   **c.** light.   **d.** mass.

_____ **15.** The convective zone of the sun is surrounded by the:

      a. corona.   **b.** photosphere.
      c. core.   **d.** chromosphere.

M O D E R N   E A R T H   S C I E N C E

Chapter 28

Choose the one best response. Write the letter of that choice in the space provided.

_____ **16.** Which of the following shields the earth from the harmful ultraviolet radiation of the sun?

    a. crust   **b.** ozone layer
    c. mantle   **d.** magnetosphere

_____ **17.** It is generally believed that the entire solar system originated from a:

    a. nebula.   **b.** protoplanet.   **c.** star.   **d.** planetesimal.

_____ **18.** Auroras are an effect caused by:

    a. nuclear fusion.   **b.** radio waves.
    c. magnetic storms.   **d.** solar nebulae.

_____ **19.** In which layer of the sun does granulation occur?

    a. core   **b.** chromosphere
    c. corona   **d.** photosphere

Use the diagram below to answer question 20.

Nuclear Fusion Reaction
(p=proton, n=neutron)

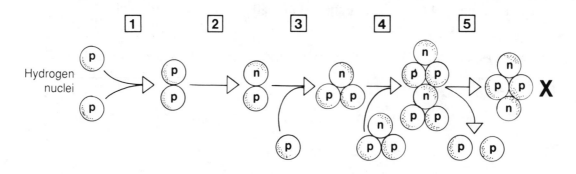

_____ **20.** Which of the following types of nuclei is substance **X**?

    a. nitrogen   **b.** iron   **c.** helium   **d.** carbon

Complete each statement by writing the correct term or phrase in the space provided.

21. The layer of the sun's atmosphere that prevents most of the atomic particles of the sun's

surface from escaping into space is the _____ .

22. The region of the sun in which energy is transferred between atoms in the form of

electromagnetic waves is the _____ .

M O D E R N   E A R T H   S C I E N C E

Chapter 28

Complete each statement by writing the correct term or phrase in the space provided.

23. Hydrogen fusion occurs within the sun's _____ .

24. As solar wind particles enter the earth's atmosphere, they can generate disturbances

    called _____ .

25. The layer considered to be the surface of the sun is the _____ .

26. When the earth cooled, oceans formed as the water vapor _____ .

27. The earth's core is composed mostly of the two elements _____ .

28. The small bodies that formed from protoplanets and that now orbit the existing planets are

    called _____ .

29. One of the first indications to astronomers that the sun rotates on its axis was the movement

    of _____ .

30. The northern and southern lights are also called _____ .

Read each question or statement and answer it in the space provided.

31. Explain how sunspots are caused by powerful magnetic fields.

    _____

    _____

    _____

    _____

    _____

32. Explain how the sun formed.

    _____

    _____

    _____

    _____

    _____

M O D E R N   E A R T H   S C I E N C E

Chapter 28

Read each question or statement and answer it in the space provided.

33. Explain how heat energy is transferred to the sun's surface by convection.

_____

_____

_____

_____

_____

34. Describe the way in which the sun rotates.

_____

_____

_____

_____

_____

35. If the earth were much more massive than it is today and had no green plants on its surface, what changes in the earth's atmosphere might be expected?

_____

_____

_____

_____

_____

Name _____ Class _____ Date _____

M O D E R N   E A R T H   S C I E N C E

Chapter 29
The Solar System

Read each statement below. If the statement is true, write _T_ in the space provided. If the statement is false, write _F_ in the space provided.

_____ **1.** Aristotle developed the first accurate model of the solar system.

_____ **2.** The geocentric model of the solar system was proposed by Copernicus.

_____ **3.** The point where an orbit comes closest to the sun is called perihelion.

_____ **4.** Hardened lava in some craters on Mercury suggests that the planet was once volcanic.

_____ **5.** The surface of Venus is probably covered with oceans.

_____ **6.** Mars has polar ice caps.

_____ **7.** Jupiter rotates faster than any other planet in the solar system.

_____ **8.** Neptune is composed largely of ice.

_____ **9.** The tail of a comet usually points toward the sun.

_____ **10.** Fireballs are produced by meteors.

Choose the one best response. Write the letter of that choice in the space provided.

_____ **11.** One astronomical unit is equal to the distance between the earth and:

      a. Mercury.   **b.** the moon.   **c.** Pluto.   **d.** the sun.

_____ **12.** The first astronomer to mathematically model most aspects of planetary motion was:

      a. Galileo.   **b.** Brahe.   **c.** Copernicus.   **d.** Kepler.

_____ **13.** Which of the following is a terrestrial planet?

      a. Pluto   **b.** Uranus   **c.** Mars   **d.** Jupiter

_____ **14.** Which of the following is hypothesized to originate in the Oort cloud?

      a. comets   **b.** fireballs   **c.** meteors   **d.** asteroids

_____ **15.** Clouds in the atmosphere of Venus are composed of droplets of:

      a. liquid hydrogen.   **b.** water vapor.
      c. carbon dioxide.   **d.** sulfuric acid.

Chapter 29

Choose the one best response. Write the letter of that choice in the space provided.

_____ 16. The largest volcanoes in the solar system are most likely on:

      a. Mars.   **b.** Jupiter.   **c.** Earth.   **d.** Pluto.

_____ 17. The first planet whose existence was predicted before it was discovered was:

      a. Venus.   **b.** Neptune.   **c.** Uranus.   **d.** Pluto.

Use the diagram below to answer questions 18-20.

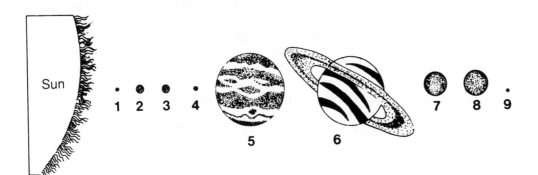

_____ 18. The bands on planet **5** are caused by:

      a. volcanic activity.   **b.** rotating moons.
      c. swirling gases.   **d.** nuclear fusion.

_____ 19. What is the name of planet **7**?

      a. Pluto   **b.** Venus   **c.** Neptune   **d.** Uranus

_____ 20. Which planet has the lowest density?

      a. 2   **b.** 4   **c.** 6   **d.** 8

Read each question and answer it in the space provided.

21. Who first used a telescope to observe celestial objects?   _____

22. What is the shape of the path followed by each planet as it orbits the sun?   _____

23. What are the bowl-shaped depressions that result from the collision of a planet with meteors?   _____

24. What is another name for the outer planets?   _____

M O D E R N   E A R T H   S C I E N C E

Chapter 29

Read each question and answer it in the space provided.

25. What is the name for the asteroids that are concentrated ahead of and behind Jupiter?

_____

26. What is visible when large numbers of small meteoroids enter the earth's atmosphere during a short time?

_____

27. Which astronomer proposed planetary motions called epicycles?

_____

28. What type of comet completes its orbit in less than 100 years?

_____

29. What U.S. spacecraft transmitted the first pictures of Uranus?

_____

30. What is a meteor called after it hits the earth?

_____

Read each question or statement and answer it in the space provided.

31. Describe the law of equal areas.

_____

_____

_____

_____

_____

_____

32. Give two explanations for Mercury's lack of an atmosphere.

_____

_____

_____

_____

_____

_____

M O D E R N   E A R T H   S C I E N C E

Chapter 29

Read each question or statement and answer it in the space provided.

33. How does the earth's distance from the sun make it more suitable for life than the other planets?

_____

_____

_____

_____

_____

_____

34. What is distinctive about the rotation of Uranus?

_____

_____

_____

_____

_____

35. Describe the effects of inertia on a moving and a stationary object.

_____

_____

_____

_____

_____

M O D E R N   E A R T H   S C I E N C E

Chapter 30
Moons and Rings

Read each statement below. If the statement is true, write _T_ in the space provided. If the statement is false, write _F_ in the space provided.

_____ **1.** The force of gravity on the moon is greater than it is on earth.

_____ **2.** Seismographs have been placed on the moon by astronauts.

_____ **3.** During the fourth and final stage of lunar development, large numbers of meteorites punctured the moon's solid crust.

_____ **4.** The moon revolves around the earth in a circular orbit.

_____ **5.** An eclipse occurs when one planetary body passes through the shadow of another.

_____ **6.** A new moon occurs when the lighted surface of the moon faces toward the earth.

_____ **7.** A planetary body makes one rotation on its axis each day.

_____ **8.** The moons of Mars are heavily cratered.

_____ **9.** Jupiter's ring is made of particles of dark rock.

_____ **10.** A promising method for mining moon rocks uses the process of hydrolysis.

Choose the one best response. Write the letter of that choice in the space provided.

_____ **11.** The streaks that form when surface material has been splashed around lunar craters are called:

       a. rilles.   **b.** maria.   **c.** regolith.   **d.** rays.

_____ **12.** Which of the following will occur if the moon crosses the plane of the earth's orbit when the moon is between the earth and the sun?

       a. solar eclipse   **b.** full moon phase
       c. lunar eclipse   **d.** last-quarter phase

_____ **13.** As the moon rotates, one side of the moon always faces the earth because:

       a. the moon rotates and revolves at the same rate.
       b. the earth's gravity attracts the near side.
       c. the moon has more mass than the earth.
       d. gravity on the moon is greater than on earth.

M O D E R N   E A R T H   S C I E N C E

Chapter 30

Choose the one best response. Write the letter of that choice in the space provided.

_____ **14.** When the visible portion of the moon is increasing, the moon is:

    a. waxing.   **b.** waning.

    c. full.   **d.** waning-crescent.

_____ **15.** What is the name of the calendar that was introduced in 46 B.C. to make 365.25 days the average length of a year?

    a. World   **b.** Sol   **c.** Gregorian   **d.** Julian

_____ **16.** Phobos is a moon of:

    a. Venus.   **b.** Mars.   **c.** Jupiter.   **d.** Saturn.

_____ **17.** Which of the following is a moon of Jupiter?

    a. Io   **b.** Titan   **c.** Oberon   **d.** Triton

_____ **18.** A body that orbits a larger body is called a:

    a. penumbra.   **b.** satellite.   **c.** ray.   **d.** regolith.

Use the diagram below to answer questions 19 and 20.

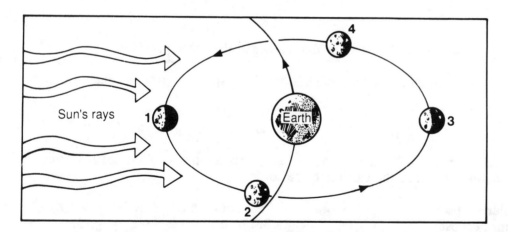

_____ **19.** A solar eclipse is most likely to occur when the moon is in position:

    a. 1.   **b.** 2.   **c.** 3.   **d.** 4.

_____ **20.** Observers on earth see a full moon when the moon is in position:

    a. 1.   **b.** 2.   **c.** 3.   **d.** 4.

M O D E R N   E A R T H   S C I E N C E

Chapter 30

Read each question and answer it in the space provided.

21. What is the average distance of the moon from the earth? _____

22. When the moon is farthest from the earth, it is said to be at what position? _____

23. What is the layer beneath the moon's crust called? _____

24. What phase is the moon in when the side of the moon facing the earth is unlighted? _____

25. How long does it take for the moon to make one revolution around the earth? _____

26. What part of a planetary body's shadow is cone shaped and completely blocks the sunlight? _____

27. What is sunlight reflected off the earth called? _____

28. In what period of time does the moon go through a complete cycle of phases? _____

29. Which planet has two small, irregularly shaped moons? _____

30. Which planet has a moon called Triton, which is known to revolve from east to west around the planet? _____

Read each question or statement and answer it in the space provided.

31. Describe the giant-impact hypothesis of the moon's origin.

_____

_____

_____

_____

_____

M O D E R N   E A R T H   S C I E N C E

Chapter 30

Read each question or statement and answer it in the space provided.

32. Describe an annular solar eclipse.

_____

_____

_____

_____

_____

33. What are the two theories used to explain the formation of Saturn's rings?

_____

_____

_____

_____

_____

34. Why is the moon's surface different from the earth's surface?

_____

_____

_____

_____

_____

35. Would an individual on the moon weigh the same, more, or less than he or she weighs on the earth? Explain your answer.

_____

_____

_____

_____

_____

M O D E R N   E A R T H   S C I E N C E

Unit 8
Studying Space

Read each statement below. If the statement is true, write *T* **in the space provided.**
If the statement is false, write *F* **in the space provided.**

_____ 1. Maria are long, deep channels found on the lunar surface.

_____ 2. Galileo's observations with a telescope helped confirm Copernicus' model of the solar system.

_____ 3. Venus is one of the Jovian planets.

_____ 4. The most common nuclear reaction occurring inside the sun is nuclear fusion.

_____ 5. The northern and southern lights are effects of magnetic storms.

_____ 6. Scientists can compare a star's apparent and absolute magnitudes to determine its distance from the earth.

_____ 7. The side of the moon facing the earth always receives the same amount of sunlight as the side facing away from the earth.

_____ 8. The moon may have formed by being struck by a Mars-sized body.

_____ 9. The planet Jupiter has only four moons.

_____ 10. The Doppler effect is commonly used to measure the absolute magnitudes of stars far away from the earth.

_____ 11. Spiral galaxies are smaller and fainter than other types of galaxies.

_____ 12. As the solid earth developed, the less-dense materials flowed to its center.

_____ 13. Neptune's upper atmosphere is composed of frozen methane.

_____ 14. The asteroid belt is located between the orbits of Mars and Jupiter.

_____ 15. A light-year is the distance that light travels in one year.

Choose the one best response. Write the letter of that choice in the space provided.

_____ 16. How many rotations does the earth make on its axis in a day?

   a. 1   **b.** 2   **c.** 3   **d.** 4

_____ 17. Kepler's law describing the shape of planetary orbits is called the law of:

   a. equal areas.   **b.** periods.   **c.** epicycles.   **d.** ellipses.

M O D E R N   E A R T H   S C I E N C E

Unit 8

Choose the one best response. Write the letter of that choice in the space provided.

_____ **18.** Which of the following planets may have lost its original atmosphere of lighter gases due to an intense solar wind?

      a. Mars   **b.** Neptune   **c.** Jupiter   **d.** Uranus

_____ **19.** The core of the earth consists primarily of nickel and:

      a. helium.   **b.** carbon.   **c.** iron.   **d.** magnesium.

Use the table below to answer questions 20 and 21.

Planetary Data

Planet	Average distance from sun (10^6 km)	Diameter (km)	Rate of rotation (Earth time)
1	57.9	4,878	59 d
2	108.2	12,104	−243 d
3	149.6	12,756	23 hr 56 min
4	227.9	6,794	24 hr 37 min
5	778.3	142,796	9 hr 50 min
6	1,427	120,660	10 hr 40 min
7	2,871	52,400	17 hr 14 min
8	4,497	50,450	16 hr 06 min
9	5,914	2,445	6.4 d

_____ **20.** Which planet in the table represents Jupiter?

      a. 4   **b.** 5   **c.** 6   **d.** 7

_____ **21.** According to Kepler's third law, which planet has the longest orbit period?

      a. 1   **b.** 2   **c.** 5   **d.** 9

_____ **22.** Astronomers label the stars in a constellation according to their:

      a. color.   **b.** apparent magnitude.
      c. size.   **d.** absolute magnitude.

_____ **23.** Which of the following occurs when one planetary body passes through the shadow of another?

      a. eclipse   **b.** umbra   **c.** apogee   **d.** perigee

_____ **24.** What is the name of the calendar currently in use in the United States?

      a. Julian   **b.** Gregorian   **c.** Sol   **d.** World

M O D E R N   E A R T H   S C I E N C E

Unit 8

Choose the one best response. Write the letter of that choice in the space provided.

_____ **25.** Olympus Mons is a large volcano on:

    a. the moon.   **b.** the earth.   **c.** Mars.   **d.** Jupiter.

_____ **26.** The outermost layer of the sun's atmosphere is the:

    a. aurora.   **b.** chromosphere.
    c. corona.   **d.** radiative zone.

_____ **27.** What phase is the moon in when the unlighted side faces the earth?

    a. crescent   **b.** gibbous   **c.** full   **d.** new

_____ **28.** When a white dwarf star no longer emits energy, it may become a:

    a. black dwarf.   **b.** nova.   **c.** neutron star.   **d.** black hole.

_____ **29.** Starlight passing through a device for separating light produces a display of colors and lines called a:

    a. barred spiral.   **b.** shift.
    c. open cluster.   **d.** spectrum.

_____ **30.** A main-sequence star maintains a stable size as long as it has an ample supply of hydrogen to fuse into:

    a. helium.   **b.** magnesium.   **c.** protactinium.   **d.** iron.

_____ **31.** Compared to surrounding areas, regions of the photosphere near strong magnetic fields are:

    a. more reactive.   **b.** cooler.
    c. less dense.   **d.** brighter.

Complete each statement by writing the correct term or phrase in the space provided.

32. An earth-centered model of the solar system is called _____ .

33. Venus' dense atmosphere may have resulted from carbon dioxide released by

_____ .

34. The planet that has the largest, brightest rings in the solar system is

_____ .

35. The solar system developed from a cloud of gas and dust called a

_____ .

M O D E R N   E A R T H   S C I E N C E

Unit 8

Complete each statement by writing the correct term or phrase in the space provided.

36. Einstein proposed that a small amount of matter could produce a large amount of

_____ .

37. The most distant objects from earth are called _____ .

38. To observers in North America, most stars appear to move around the star called

_____ .

39. The planet that has two small, irregularly shaped moons that are heavily cratered is

_____ .

40. Light-colored, coarse-grained rocks called anorthosites make up much of the lunar surface

called the _____ .

41. Stars that are smaller than neutron stars that contract with great force, crushing their cores

and leaving gaps in space are called _____ .

42. In the radiative zone, heat energy moves from atom to atom in the form of

_____ .

43. Clouds of glowing gases that follow curved lines of magnetic force above the sun's surface

are called _____ .

44. The rarest of the three basic types of meteorites is called _____ .

Read each question or statement and answer it in the space provided.

45. What two proposals were stated in Laplace's hypothesis on the origin of the solar system?

_____

_____

_____

_____

_____

_____

_____

M O D E R N   E A R T H   S C I E N C E

Unit 8

Read each question or statement and answer it in the space provided.

46. Identify the three parts of a comet and describe their composition.

_____

_____

_____

_____

_____

_____

47. Explain how parallax can be used to measure the distance between the earth and a star.

_____

_____

_____

_____

_____

_____

48. Explain how a nova occurs.

_____

_____

_____

_____

_____

_____

M O D E R N   E A R T H   S C I E N C E

Unit 8

Read each question or statement and answer it in the space provided.

49. How does the rate of lunar rotation affect the way we see the moon?

_____

_____

_____

_____

_____

_____

50. What three sources of heat contributed to the high temperature on the earth as it formed?

_____

_____

_____

_____

_____

_____

M O D E R N   E A R T H   S C I E N C E

Final Test
Earth Science

Read each statement below. If the statement is true, write *T* in the space provided. If the statement is false, write *F* in the space provided.

_____ 1. Ecology is a field of study that unites earth science and biology.

_____ 2. Galaxies in the universe are moving closer together.

_____ 3. The movement of magma toward or onto the earth's surface is called volcanism.

_____ 4. An area in which one lithospheric plate is being moved under another is called a subduction zone.

_____ 5. Ionic and covalent are types of bonds.

_____ 6. An isotope is a type of subatomic particle.

_____ 7. It is possible for a porous rock to be impermeable.

_____ 8. An artesian formation is bounded by two layers of impermeable rock.

_____ 9. Pangaea was a large supercontinent that formed in Precambrian time.

_____ 10. Sediments eroded from the Sierra Nevada eventually formed the Great Plains.

_____ 11. Waves break primarily as a result of friction.

_____ 12. The difference in level between high and low tide at a particular location is called the tidal bore.

_____ 13. Latitude determines both the amount of solar energy an area receives and the prevailing wind patterns of the region.

_____ 14. Middle-latitude deserts have seasonal temperature ranges similar to those of tropical deserts.

_____ 15. Lunar surface features have undergone significant modification since their formation.

_____ 16. Eclipses are the result of one planetary body casting a shadow on another.

_____ 17. From Earth, stars appear to rise in the west and set in the east.

_____ 18. The galaxy that includes the earth is called the Large Magellanic Cloud.

Choose the one best response. Write the letter of that choice in the space provided.

_____ 19. A possible explanation of or solution to a problem is called:

    a. a law.   **b.** a hypothesis.
    c. a conclusion.   **d.** an observation.

M O D E R N   E A R T H   S C I E N C E

Final Test

Choose the one best response. Write the letter of that choice in the space provided.

_____ **20.** The first step in using scientific methods to solve a problem is to:

   a. state the problem.   **b.** form a hypothesis.
   c. reach a conclusion.   **d.** gather information.

_____ **21.** The lithosphere is made up of the upper mantle and the:

   a. crust.   **b.** asthenosphere.
   c. hydrosphere.   **d.** core.

_____ **22.** When the South Pole tilts toward the sun, the Northern Hemisphere experiences:

   a. fall.   **b.** winter.   **c.** spring.   **d.** summer.

_____ **23.** A satellite that passes over a different portion of the earth during each revolution is traveling:

   a. at apogee.   **b.** in a geosynchronous orbit.
   c. at perigee.   **d.** in a polar orbit.

_____ **24.** What type of boundary forms when two lithospheric plates collide?

   a. rift valley   **b.** divergent
   c. transform fault   **d.** convergent

_____ **25.** Pieces of land bounded by faults that have different geologic features from those of neighboring land are most likely to be:

   a. island arcs.   **b.** hot spots.
   c. suspect terranes.   **d.** spreading centers.

_____ **26.** Which of the following is the major cause of deformation of the earth's crust?

   a. magma formation   **b.** hot spots
   c. plate tectonics   **d.** divergent margins

_____ **27.** The permanent deformation of a rock without the rock breaking is called:

   a. collision.   **b.** folding.   **c.** faulting.   **d.** fracturing.

_____ **28.** Which of the following usually form along convergent plate boundaries?

   a. mountains   **b.** normal faults   **c.** grabens   **d.** ocean basins

_____ **29.** Geologists use the elastic rebound theory to explain:

   a. the cause of tsunamis.
   b. the intensity of an earthquake.
   c. the magnitude of tsunamis.
   d. the cause of an earthquake.

Final Test

Choose the one best response. Write the letter of that choice in the space provided.

_____ **30.** Which of the following best describes aftershocks?

    a. a series of small tremors occurring after a major earthquake
    b. seismic waves that cannot travel through liquids
    c. areas along a fault where slippage and fracturing first occur
    d. giant ocean waves that originate at a fault zone

_____ **31.** How many times more energy is released by an earthquake having a magnitude of 3.5 than by an earthquake having a magnitude of 2.5?

    a. 1,000 times more energy   **b.** 317 times more energy
    c. 100 times more energy   **d.** 31.7 times more energy

_____ **32.** The material that collects around the vent of a volcano as a result of eruptions are classified into three major types of volcanic:

    a. fissures.   **b.** calderas.   **c.** craters.   **d.** cones.

_____ **33.** In order to be classified as a mineral, a substance must be:

    a. organic.   **b.** a solid.   **c.** a metal.   **d.** crystalline.

_____ **34.** Which of the following tests indicates a mineral's color in powdered form?

    a. hardness   **b.** luster   **c.** streak   **d.** fracture

_____ **35.** Which of the following is a nonsilicate mineral?

    a. halite   **b.** quartz   **c.** feldspar   **d.** talc

_____ **36.** Which of the following is an igneous rock?

    a. limestone   **b.** gypsum   **c.** gneiss   **d.** basalt

_____ **37.** Regional metamorphism occurs as a result of:

    a. tectonic activity.   **b.** volcanic eruptions.
    c. earthquakes.   **d.** sedimentation.

_____ **38.** To which of the following rock types does breccia belong?

    a. foliated metamorphic   **b.** unfoliated metamorphic
    c. clastic sedimentary   **d.** chemical sedimentary

_____ **39.** Which of the following describes the process by which sedimentary rock becomes metamorphic rock?

    a. weathering   **b.** erosion
    c. intense heat and pressure   **d.** cooling and solidifying

_____ **40.** The hardest form of coal is:

    a. peat.   **b.** anthracite.
    c. lignite.   **d.** bituminous coal.

M O D E R N   E A R T H   S C I E N C E

Final Test

Choose the one best response. Write the letter of that choice in the space provided.

_____ **41.** Placer deposits form at the bottom of:

    a. stream beds.   **b.** oceans.
    c. mountains.   **d.** magma bodies.

_____ **42.** The fuel for nuclear fission reactors is:

    a. plutonium.   **b.** uranium.   **c.** hydrogen.   **d.** fossil fuels.

_____ **43.** The energy source for geothermal energy is:

    a. magma.   **b.** the sun.
    c. wind.   **d.** radioactive ores.

_____ **44.** Which of the following is usually the slowest type of mass movement?

    a. landslide   **b.** slump   **c.** rockfall   **d.** creep

_____ **45.** Rusting is an example of which of the following processes?

    a. hydrolysis   **b.** oxidation   **c.** exfoliation   **d.** carbonation

_____ **46.** Till is best defined as:

    a. an unsorted deposit of rock material.
    b. sorted and layered deposits of sand.
    c. sediment sorted by melted ice.
    d. unsorted fan-shaped deposits of drift.

_____ **47.** Cirques are best described as:

    a. rounded knobs of rock.   **b.** sharp, curved peaks.
    c. bowl-shaped depressions.   **d.** sharp and jagged ridges.

_____ **48.** The most accurate method of finding the absolute age of a rock is by:

    a. varve count dating.   **b.** erosion rate dating.
    c. radioactive decay dating.   **d.** superposition dating.

_____ **49.** The ordered arrangement of rock layers is called the:

    a. geologic column.   **b.** radioactive dating.
    c. theory of evolution.   **d.** geologic time scale.

_____ **50.** During what era did mammals become the predominant life form?

    a. Precambrian   **b.** Paleozoic   **c.** Mesozoic   **d.** Cenozoic

_____ **51.** Which ocean zone is characterized by breaking waves?

    a. neritic   **b.** benthic   **c.** intertidal   **d.** abyssal

_____ **52.** Which ocean zone separates warm surface water from the colder deep water?

    a. thermocline   **b.** benthos   **c.** abyssal plain   **d.** neritic zone

M O D E R N   E A R T H   S C I E N C E

Final Test

Choose the one best response. Write the letter of that choice in the space provided.

_____ 53. At what time does the highest temperature of the day typically occur?

      a. 10:00 A.M.   **b.** 12:00 noon   **c.** 3:00 P.M.   **d.** 6:00 P.M.

_____ 54. The atmospheric layer closest to the earth's surface is called the:

      a. troposphere.   **b.** stratosphere.   **c.** ionosphere.   **d.** mesosphere.

_____ 55. In what region do the northeast and southeast trade winds meet?

      a. horse latitudes   **b.** doldrums
      c. subpolar lows   **d.** jet streams

_____ 56. The mass of water vapor in a sample of air, compared to the mass of water vapor the air can hold at saturation, is called the:

      a. dew point.   **b.** relative humidity.
      c. condensation level.   **d.** specific humidity.

_____ 57. Which of the following cloud types occurs at the highest altitude?

      a. cumulus   **b.** stratus   **c.** cirrus   **d.** nimbus

_____ 58. Precipitation consisting of drops smaller than 0.5 mm in diameter is called:

      a. drizzle.   **b.** rain.   **c.** sleet.   **d.** snow.

_____ 59. What is the general name for a large body of air having uniform temperature and moisture content?

      a. wind cell   **b.** air mass
      c. stationary front   **d.** wave cyclone

_____ 60. What instrument measures wind direction?

      a. wind scale   **b.** wind compass
      c. wind gauge   **d.** wind vane

_____ 61. On a weather map, station models are used to indicate weather conditions:

      a. near an observation center.   **b.** around a storm system.
      c. over the oceans.   **d.** in the upper atmosphere.

_____ 62. Cooler regions of the photosphere near strong magnetic fields are called:

      a. solar flares.   **b.** prominences.   **c.** solar winds.   **d.** sunspots.

_____ 63. The surface of the sun is called the:

      a. troposphere.   **b.** corona.   **c.** photosphere.   **d.** aurora.

_____ 64. Who first proposed a heliocentric model of the solar system?

      a. Ptolemy   **b.** Copernicus   **c.** Brahe   **d.** Galileo

M O D E R N   E A R T H   S C I E N C E

Final Test

Choose the one best response. Write the letter of that choice in the space provided.

_____ 65. The terrestrial planets are composed mostly of:

a. liquid metal.   b. ice.   c. solid rock.   d. gases.

_____ 66. The most common form of wind erosion is called:

a. saltation.   b. deflation.   c. hollowing.   d. deposition.

_____ 67. Erosion along shorelines is primarily caused by:

a. salt.   b. currents.   c. waves.   d. wind.

_____ 68. Many emergent and submergent coastline features are formed by changes in:

a. sea level.   b. wind direction.
c. atmospheric pressure.   d. chemical weathering.

_____ 69. The small bodies of matter that were present in the solar nebula while the sun was forming are called:

a. auroras.   b. planetesimals.
c. moons.   d. protoplanets.

Use the diagram below to answer questions 70 and 71.

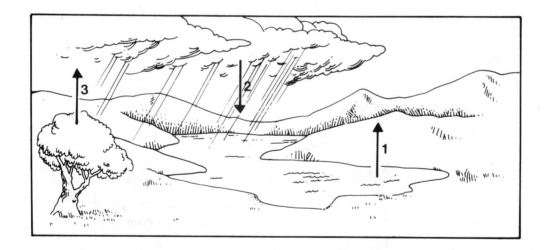

_____ 70. The process that must occur before the process labeled **2** can occur is called:

a. transpiration.   b. precipitation.
c. condensation.   d. evaporation.

_____ 71. The process represented by arrow **3** is:

a. transpiration.   b. precipitation.
c. condensation.   d. evaporation.

M O D E R N   E A R T H   S C I E N C E

Final Test

Complete each statement by writing the correct term or phrase in the space provided.

72. Angular distance east and west of the prime meridian is called _____ .

73. The map feature that indicates the relationship between distance as it is drawn on the map and actual distance is called the _____ .

74. The type of map that shows the elevation of the surface features of the earth is called a

_____ .

75. The theory proposing that continents were once joined in a single landmass is called the

theory of _____ .

76. A substance that cannot be broken down into a simpler form by ordinary chemical means is

called _____ .

77. The principle that states that each sedimentary rock layer is older than the layers above it and younger than the layers below it is called the _____ .

78. Scientists use index fossils to determine a rock's _____ .

79. Breaks in the geologic record that make it difficult to determine the relative age of rock

layers are called _____ .

80. The periods of the Cenozoic Era have been divided into smaller time spans called

_____ .

81. Animals without backbones are called _____ .

82. Parts of the ocean floor are covered with a soft, organic sediment called

_____ .

83. The portion of the continental margin that is covered by the shallowest water is the

_____ .

84. The technique used in underwater oceanographic research that makes use of sound waves is

_____ .

M O D E R N   E A R T H   S C I E N C E

Final Test

Complete each statement by writing the correct term or phrase in the space provided.

85. The type of air mass that moves over land is called a _____ .

86. The deflection of wind caused by the earth's rotation is called the

_____ .

87. The true brightness of a star is called its _____ .

88. The photosphere, chromosphere, and corona are the three layers of the sun's

_____ .

89. The primary element fueling the sun's nuclear fusion reactions is

_____ .

90. Planets are kept in their orbits by the gravitational attraction of the

_____ .

91. The head of a comet is made up of the coma and the _____ .

92. Pedocal soils are soils that contain large amounts of the mineral

_____ .

93. Periods during an ice age when glaciers advance are called _____ .

94. Most outwash plains are pitted with depressions called _____ .

95. Almost two thirds of the earth's mass is contained in the layer called the

_____ .

Read each question or statement and answer it in the space provided.

96. How does a large lake, such as Lake Michigan, affect climate?

_____

_____

_____

_____

M O D E R N   E A R T H   S C I E N C E

Final Test

Read each question or statement and answer it in the space provided.

97. Describe the structure of the moon.

_____

_____

_____

_____

98. Describe the Canadian Shield and its location.

_____

_____

_____

_____

99. Explain what is meant by the term mixture.

_____

_____

_____

_____

100. Describe how coral reefs are formed.

_____

_____

_____

_____

M O D E R N E A R T H S C I E N C E

Answers to Safety Test

| 1. F | 2. T | 3. T | 4. F | 5. F | 6. c | 7. b | 8. d | 9. d | 10. c |

Answers to Metric Test

| 1. F | 2. F | 3. T | 4. F | 5. T | 6. a | 7. d | 8. d | 9. c | 10. b |

M O D E R N   E A R T H   S C I E N C E

Answers to Chapter 1 Test

1. F	2. F	3. T	4. T	5. T	6. F	7. T	8. T	9. F	10. T
11. b	12. a	13. b	14. c	15. c	16. a	17. c	18. d	19. c	20. d

21. earth science
22. atmosphere
23. pollution
24. ozone
25. control
26. scientific methods
27. hypothesis
28. spectrum
29. element
30. shift

31. Plastic waste dumped into oceans and lakes harms life there. Ingested particles of plastic can clog the digestive tracts of fish and other animals. Animals can also starve to death when they become tangled in plastic litter and become unable to move.

32. Billions of years ago, all matter and energy in the universe was compressed into an extremely small volume that suddenly expanded, sending the matter and energy hurtling outward in a giant cloud. As the cloud expanded, some of the matter condensed into galaxies. Today the universe is still expanding.

33. Plants in the rain forest use sunlight to produce food. The plants are then eaten by animals, which in turn are eaten by other animals. When the plants and animals die, microorganisms decompose them. The chemicals released during decomposition enter the soil and eventually nourish other plants and animals.

34. Scientists have found that the bright-line spectrum of nearly every galaxy tested is shifted toward the red end while only a few close galaxies are blue-shifted. The red shift indicates that almost all of the galaxies in the universe are moving away from the earth, and that the universe is expanding.

35. Space travelers could monitor the bright-line spectrum emitted by the galaxy. If the spectrum is shifted toward the red end, the travelers are moving away from the galaxy. If the spectrum is shifted toward the blue end, the travelers are moving toward the galaxy.

Guide to Tested Objectives

Objective **1**: questions 1(A), 10(A), 14(A), 21(A); Objective **2**: questions 2(A), 3(A), 12(A), 22(A), 23(A), 24(A), 31(A), 33(B); Objective **3**: questions 4(B), 11(A), 13(A)-feature, 25(A), 26(A); Objective **4**: questions 5(A), 7(A), 18(B); Objective **5**: questions 8(A), 16(A), 27(A); Objective **6**: questions 9(A), 19(B), 20(B), 28(A), 29(A), 35(B); Objective **7**: questions 6(A), 30(B), 32(A); Objective **8**: questions 15(A), 17(A), 34(B)

Answers to Chapter 2 Test

1. T	2. T	3. T	4. F	5. F	6. T	7. T	8. T	9. T	10. F
11. a	12. b	13. a	14. b	15. c	16. d	17. b	18. a	19. b	20. c

21. solstice
22. Moho
23. lithosphere
24. weight
25. perihelion
26. precession
27. daylight saving time
28. clock
29. geosynchronous
30. perigee

31. The motion of the earth's rotation causes the earth to bulge slightly at the equator and flatten at the poles. The distance between the surface and the earth's center is, therefore, greatest at the equator and least at the poles. Since the force of gravity decreases as distance from the earth's center increases, the weight of an object will be more at the poles than at the equator.

32. When the North Pole tilts toward the sun, the Northern Hemisphere has longer periods of daylight than the Southern Hemisphere. When the North Pole tilts away from the sun, the Southern Hemisphere has longer periods of daylight. On the autumnal equinox and on the vernal equinox, the North Pole tilts neither toward nor away from the sun, and on those days, hours of daylight and darkness are equal everywhere on earth.

33. When the North Pole tilts away from the sun, the angle of the sun's rays on the Northern Hemisphere decreases and the sun's rays spread out. The weak rays of the sun and shorter daylight hours cause the winter season. When the North Pole tilts toward the sun in summer, the sun's rays strike the earth at higher angles. The increased hours of daylight and the sun's more direct rays cause the summer season.

34. Newton's law of gravitation states that the force of attraction between any two objects depends upon their masses and the distance between them. The larger the masses of two objects and the closer together they are, the greater will be the force of gravity between them.

35. Nome is on the east side of the international date line and Sydney is on the west side, because Sydney is one day ahead of Nome. Also, Sydney is two time zones west of Nome because the time in Sydney is two hours earlier.

Guide to Tested Objectives

Objective **1**: questions 1(A), 2(A), 17(A), 23(A); Objective **2**: questions 3(A), 16(B), 22(A); Objective **3**: questions 4(A), 11(A); Objective **4**: questions 5(B), 12(B), 24(A), 31(A), 34(B); Objective **5**: questions 6(A), 13(A), 15(B), 25(A), 32(A); Objective **6**: questions 7(A), 21(B), 26(A), 33(A); Objective **7**: questions 8(A), 19(A), 20(B), 27(A), 28(A), 35(B); Objective **8**: questions 9(A), 14(A), 29(A), 30(A); Objective **9**: questions 10(B), 18(B)-feature

HRW material copyrighted under notice appearing earlier in this work.

195

M O D E R N   E A R T H   S C I E N C E

Answers to Chapter 3 Test

1. F	2. F	3. T	4. F	5. T	6. T	7. T	8. F	9. F	10. T
11. c	12. b	13. c	14. d	15. d	16. b	17. a	18. a	19. c	20. b

21. prime meridian
22. latitude
23. declination
24. great circle
25. contour interval
26. polyconic projections
27. legend
28. equator
29. index contours
30. depression contours

31. All lines of longitude intersect at the poles. As these lines approach the equator, they are farther and farther apart. Therefore, the distance between two adjacent lines of longitude, which represents one degree, is greater at the equator than at 45° N latitude.

32. Topographic maps show the surface features, or topography, of an area. Positions of hills, valleys, and rivers, and steepness of slopes can be determined from such maps. This type of information would be useful in making decisions about a route to be hiked.

33. Magnetic compasses point toward the geomagnetic North Pole. In order to determine true geographic north using a magnetic compass, the magnetic declination must be known. By correcting the compass reading to account for the declination, geographic north may be determined.

34. Navigators can use great circles to show the shortest distance between any two points on the earth. On gnomonic projections, great circles appear as straight lines.

35. The city has a latitude of 20 degrees south. Therefore, it is located south of the equator (i.e., in the southern hemisphere). The city's longitude is 50 degrees west. Therefore, it is also located 50 degrees west of the prime meridian.

Guide to Tested Objectives

Objective 1: questions 1(A), 10(A), 21(A), 28(A); Objective 2: questions 2(A), 3(B), 8(A), 11(A), 15(A), 18(A), 22(A), 24(A), 31(B), 35(B); Objective 3: questions 9(A), 23(A), 33(A); Objective 4: questions 4(A), 12(A), 17(B), 26(A), 27(A), 34(B); Objective 5: question 13(B); Objective 6: questions 6(A), 7(A), 14(B), 16(A), 25(A), 30(A), 32(A); Objective 7: questions 5(A)-feature, 19(B), 20(B), 29(A)

Answers to Unit 1 Test

1. T	2. F	3. T	4. T	5. F	6. F	7. T	8. F	9. T	10. T
11. F	12. T	13. T	14. F	15. T	16. b	17. d	18. c	19. b	20. b
21. b	22. b	23. a	24. c	25. d	26. c	27. d	28. a		

29. biodegradable
30. observation
31. 90°
32. seismic wave
33. daylight saving time
34. law
35. mean sea level
36. variable
37. equator
38. shorter
39. vernal equinox
40. International Date Line
41. Mercator projection
42. prime meridian
43. precession
44. shadow zones
45. 1:10,000

46. On a gnomonic projection, the unequal spacing between parallels causes great distortion in both direction and distance. The distortion increases as the distance from the point of contact used to make the projection increases. Areas on a conic projection along the parallel of latitude where the cone and globe are in contact are distorted least.

47. By December, the earth is halfway through its orbit and the North Pole tilts away from the sun. On the day of the winter solstice (December 21 or 22), the sun's rays strike the earth at a 90° angle along the Tropic of Capricorn.

48. Because of the tilt of the imaginary magnet inside the earth, the geomagnetic poles and the geographic poles are located in different places. A compass needle points to the geomagnetic pole. The angle between the direction of the geographic pole and the direction in which the compass needle points is called magnetic declination.

49. The thin, solid outermost zone of the earth is the crust. The crust is composed of oceanic and continental crust. Beneath the crust is the mantle. The uppermost part of the mantle is solid; the rest of the mantle is a solid that can flow. Below the mantle is the core, which contains iron. The outer core is liquid; the inner core is solid.

50. This hypothesis states that millions of years ago a giant meteorite crashed into the earth. The impact of the collision raised dust that blocked the sun's rays and probably caused the earth to become colder. Plant life began to die, and many animal species, including the dinosaurs, became extinct.

Guide to Tested Objectives

Chapter 1—Objective 1: questions 13(A), 21(A); Objective 2: questions 18(A), 29(A); Objective 3: questions 6(B), 12(A), 26(A), 30(A), 36(A); Objective 4: questions 2(B), 50(A); Objective 5: question 34(A); Objective 6: questions 1(A), 38(B); Objective 7: question 15(A); Objective 8: question 17(A). **Chapter 2**—Objective 1: questions 4(A), 24(A), 49(B); Objective 2: questions 5(A), 32(A), 44(A); Objective 3: question 19(A); Objective 4: question 9(A); Objective 5: questions 25(A), 43(A); Objective 6: questions 10(A), 39(A), 47(B); Objective 7: questions 22(A), 33(A), 40(A); Objective 8: question 14(A); Objective 9: questions 23(B). **Chapter 3**—Objective 1: questions 7(A), 37(A), 42(A); Objective 2: questions 8(A), 16(A), 31(B); Objective 3: question 48(A); Objective 4: questions 3(A), 28(A), 41(A), 46(B); Objective 5: question 45(B); Objective 6: questions 11(B), 35(A); Objective 7: 20(B), 27(A).

M O D E R N   E A R T H   S C I E N C E

Answers to Chapter 4 Test

1. F	2. T	3. F	4. F	5. T	6. F	7. T	8. T	9. F	10. T
11. a	12. b	13. d	14. b	15. b	16. a	17. b	18. d	19. b	20. b

21. Africa
22. theory of suspect terranes
23. Mid-Atlantic Ridge
24. magnetic orientation
25. oceanic crust and continental crust
26. terrane
27. asthenosphere
28. convection currents
29. Pangaea
30. divergent plate boundary

31. Wegener first proposed the theory of continental drift. There is evidence to support his hypothesis, such as the similarities in the coastlines of continents, the fossil remains in South America and Africa, the ages and types of rocks found on different continents, similarities among mountain chains on different continents, and the climatic histories of continents.

32. According to the theory of seafloor spreading, the tectonic plates on either side of mid-ocean ridges are moving away from each other. As the ocean floor moves away from the ridge, it is replaced by magma that wells up through the rift in the center of the ridge. This magma cools and solidifies into new ocean floor.

33. Convection currents occur beneath the earth's surface in the asthenosphere. Hot material deep within the earth's asthenosphere rises to the base of the lithosphere. Once it reaches the lithosphere, the hot material cools. New rising hot material pushes the cooler material aside. This cooler, more dense material sinks back down. This process results in a continuous circulation of material in the asthenosphere.

34. First, a terrane contains rock and fossils that differ from the rock and fossils of neighboring terranes. Second, there are major faults at the boundaries of a terrane. Finally, the magnetic properties of a terrane do not match those of neighboring terranes.

35. The oceanic plate, which is denser, is moved under the continental crust, forming a subduction zone. As it is subducted, material from the oceanic plate melts, then rises to the surface on the continental crust, forming volcanic mountains (Andes).

Guide to Tested Objectives

Objective **1**: questions 1(A), 21(B), 29(B), 31(A); Objective **2**: questions 2(B), 10(A), 24(A); Objective **3**: questions 3(B), 12(B), 13(A), 15(A), 23(A)-feature, 32(A); Objective **4**: questions 4(A), 17(A), 18(B), 19(A), 20(A), 25(A), 35(B); Objective **5**: questions 5(A), 6(A), 7(A), 11(A), 14(A), 30(B); Objective **6**: questions 8(A), 22(B), 27(B), 28(A), 33(A), 34(B); Objective **7**: questions 9(A), 16(B), 26(A)

Answers to Chapter 5 Test

1. F	2. T	3. T	4. T	5. F	6. T	7. F	8. T	9. T	10. F
11. d	12. b	13. c	14. a	15. c	16. b	17. a	18. c	19. d	20. a

21. circum-Pacific and Eurasian-Melanesian
22. faulting
23. dome mountain
24. mountain system
25. isostasy
26. compression
27. deformation
28. strike-slip fault
29. volcanic mountain
30. fault-block mountains

31. Rivers carry large amounts of sand, gravel, and rock into the oceans. When these materials are deposited in thick layers on the ocean floor, they cause the ocean floor to sink.

32. In a reverse fault, the fault plane can be almost vertical. This type of fault forms when compression causes the hanging wall to move up relative to the footwall.

33. Mount St. Helens is a volcanic mountain. Volcanic mountains form when molten lava erupts onto the surface of the earth. The lava hardens and builds up, forming mountains.

34. The oceanic crust of one plate subducts beneath the oceanic crust of the other plate. As the underlying oceanic crust plunges deeper into the mantle, the intense heat melts its crustal material, forming magma. The magma rises and breaks through the oceanic crust of the overriding plate and forms volcanic mountains on the ocean floor.

35. The San Andreas fault is a strike-slip or transform fault. These faults occur at boundaries where one plate is sliding past another and causing the rocks on either side of the fault plane to move horizontally. Shearing is the type of stress that is most likely acting on the rocks along the fault because shearing pushes rocks in two opposite horizontal directions.

Guide to Tested Objectives

Objective **1**: questions 4(A), 9(A), 12(A), 25(A); Objective **2**: questions 3(A), 5(A), 11(A), 13(B)-feature, 26(A), 27(A), 31(A); Objective **3**: questions 7(A), 8(A), 15(A), 22(A); Objective **4**: questions 10(A), 18(B), 19(B), 20(B), 28(A), 32(A), 35(B); Objective **5**: questions 2(A), 6(A), 16(A), 21(A), 24(A), 34(B); Objective **6**: questions 1(A), 14(A), 17(A), 23(A), 29(A), 30(A), 33(A)

M O D E R N   E A R T H   S C I E N C E

Answers to Chapter 6 Test

1. T	2. T	3. F	4. T	5. T	6. F	7. F	8. F	9. T	10. T
11. c	12. d	13. d	14. c	15. b	16. a	17. c	18. b	19. a	20. d

21. crust	22. P waves	23. L waves	24. seven
25. seismic gaps	26. locked	27. aftershocks	28. Pacific Ring of Fire
29. vertical motion	30. magnitude		

31. At the Mid-Atlantic Ridge, oceanic crust is pulling away from the sides of the ridge. This spreading motion creates strain in the rocks along the ridge, leading to earthquakes.

32. The Richter scale is a numerical scale that expresses the magnitude of an earthquake. The Mercalli scale uses Roman numerals and word descriptions to express the intensity of an earthquake.

33. Faulting may cause a sudden drop or rise in the ocean floor, which may result in a similar drop or rise in a mass of water. This mass of water churns up and down as it adjusts to the change in sea level and sets in motion a series of long, low waves that develop into tsunamis.

34. The point along a fault where rock first slips is called the focus. The focus may be hundreds of kilometers below the earth's surface. The epicenter is the point on the earth's surface directly above the focus.

35. You can conclude that the focus of the earthquake is far away. P waves travel 1.7 times faster than S waves. Therefore, the greater the time interval between the arrival of the two wave types, the farther away the focus is from the seismograph.

Guide to Tested Objectives

Objective **1**: questions 11(B), 21(A), 26(A), 27(A), 34(A); Objective **2**: questions 1(A), 3(A)-feature, 13(B), 28(A), 31(A); Objective **3**: questions 6(A), 12(B), 15(B), 22(A), 23(A), 35(B); Objective **4**: questions 7(A), 14(A); Objective **5**: questions 5(A), 17(A), 24(A), 29(A), 30(A), 32(A); Objective **6**: questions 2(A), 9(A); Objective **7**: questions 8(A), 18(A), 33(B); Objective **8**: questions 19(A), 20(B); Objective **9**: questions 4(A), 10(A), 16(A), 25(A)

Answers to Chapter 7 Test

1. T	2. F	3. T	4. T	5. F	6. T	7. T	8. F	9. T	10. T
11. c	12. b	13. a	14. c	15. a	16. d	17. b	18. d	19. b	20. d

21. volcanic blocks	22. hot spots	23. silica	24. shield cone
25. pahoehoe	26. mafic lava	27. composite cone	28. crater
29. caldera	30. magma		

31. When a plate with oceanic crust meets a plate with continental crust, the oceanic crust, which is the denser of the two, subducts beneath the continental crust. The plate with the continental crust buckles and folds, forming a line of mountains along the edge of the continent. Also, magma from the subducted plate rises through the continental crust, forming volcanic mountains.

32. Whether Martian volcanoes are active is a question scientists have yet to answer. However, Mars does seem to be seismically active. A Viking landing craft searching Mars for signs of life detected two geological events that produced waves similar to those of an earthquake. These "marsquakes" may also mean that the Red Planet is still volcanically active.

33. Scientists believe that Jupiter's gravitational pull moves Io inward and outward in its orbit. This pulling back and forth causes the surface of Io to move in and out. Friction from this motion probably makes the inside of Io heat up, leading to melting, magma formation, and then volcanism.

34. At mid-ocean ridges, fractures between the spreading plates along the ridges reach down to the asthenosphere. Magma rises through these fractures and comes to the surface through rifts. The emerging lava then cools and forms new ocean floor at the mid-ocean ridges.

35. The areas of volcanism within plates called hot spots remain stationary as the lithospheric plates above them continue to move. As the volcanoes that have formed are carried away from the hot spot, they cease activity because there is no longer any magma underneath to feed them. New volcanoes, however, may form where new crust has moved over the hot spot. These new volcanoes may form additional islands in the Hawaiian chain.

Guide to Tested Objectives

Objective **1**: questions 6(B), 17(B), 30(A); Objective **2**: question 3(A); Objective **3**: questions 7(A), 18(A), 22(A), 31(B), 34(B), 35(B); Objective **4**: questions 12(A), 23(B), 25(A), 26(A); Objective **5**: questions 4(A), 11(B), 21(A); Objective **6**: questions 2(A), 5(A), 10(A), 13(A), 19(B), 20(B), 27(A), 28(A), 29(A); Objective **7**: questions 1(A), 14(A), 16(A); Objective **8**: questions 9(A), 15(B), 24(B), 32(B); Objective **9**: questions 8(A), 33(B)

M O D E R N   E A R T H   S C I E N C E

Answers to Unit 2 Test

1. T	2. T	3. F	4. T	5. F	6. T	7. T	8. T	9. F	10. T
11. T	12. F	13. T	14. F	15. b	16. a	17. c	18. c	19. c	20. a
21. d	22. b	23. a	24. d	25. c	26. d	27. a	28. c	29. c	30. c

31. fault zone
32. epicenter
33. trench
34. seafloor spreading
35. terranes
36. pillow lava
37. volcanic blocks
38. volcanism
39. rift valleys
40. subduction zone
41. convection
42. asthenosphere
43. monocline
44. sink

45. A normal fault is one in which the break in the rock is almost vertical. The rocks above the normal fault plane, called the hanging wall, move down relative to the rocks below the fault plane, called the footwall.

46. Small earthquakes result from the growing pressure on the surrounding rocks as magma in the earth's mantle works its way upward. Temperature changes within the rock and the actual fracturing of the rock surrounding a volcano also contribute to small earthquakes. The number of earthquakes often increases until they occur almost continuously before a volcanic eruption.

47. The oceanic and continental crust of the earth and part of the earth's upper mantle make up the lithosphere, the solid, relatively rigid outer shell of the earth. The lithosphere is broken into separate plates. These plates ride on the denser asthenosphere beneath, a layer of solid rock that flows slowly when under pressure. The continents are carried along on the moving lithospheric plates.

48. Earthquakes usually occur when rocks that have fractured under stress suddenly shift along a fault. The rocks slide past each other, causing the trembling and vibration of an earthquake.

49. The composition of the lava that reaches the surface generally determines the force with which a particular volcano will erupt. Oceanic volcanoes are usually produced by mafic lava. Because gases can easily escape from mafic lava, eruptions of oceanic volcanoes are usually quieter than eruptions of continental volcanoes, which produce felsic lava. Felsic lavas contain a large quantity of gases that boil out of a volcano explosively when a vent or fissure opens up.

50. Two events that may cause tsunamis are faulting and underwater landslides. Faulting may bring about a sudden rise or drop in the level of a part of the ocean floor, causing a large mass of water to churn up and down. This water movement sets into motion long, low waves that develop into tsunamis. An earthquake may trigger a severe underwater landslide, throwing water above the landslide in an up-and-down motion and thereby creating a series of tsunamis.

Guide to Tested Objectives

Chapter 4—Objective 2: question 26(B); Objective 3: questions 9(B), 34(A); Objective 4: questions 7(A), 8(A), 42(A), 47(B); Objective 5: questions 6(A), 28(A), 39(A), 40(A); Objective 6: questions 27(A), 41(B); Objective 7: question 35(A).
Chapter 5—Objective 1: questions 15(A), 44(B); Objective 2: questions 1(A), 4(A), 16(A); Objective 3: question 43(A); Objective 4: questions 18(A), 45(B); Objective 5: questions 2(A), 17(A), 20(A); Objective 6: questions 5(A), 19(B).
Chapter 6—Objective 1: questions 23(B), 48(A); Objective 2: questions 11(A), 25(A), 31(A); Objective 3: questions 14(A), 22(A); Objective 4: questions 12(A), 32(A); Objective 7: question 50(B); Objective 9: question 29(B). **Chapter 7**—Objective 1: question 21(B); Objective 2: questions 24(B), 38(A); Objective 3: questions 13(A), 30(B), 33(A), 49(B); Objective 4: question 36(A); Objective 5: question 37(A); Objective 6: question 3(A); Objective 7: question 46(B); Objective 8: question 10(A).

HRW material copyrighted under notice appearing earlier in this work.

199

M O D E R N   E A R T H   S C I E N C E

Answers to Chapter 8 Test

1. F	2. F	3. T	4. T	5. F	6. F	7. T	8. F	9. T	10. T
11. c	12. c	13. a	14. c	15. d	16. a	17. a	18. b	19. b	20. b

21. elements   22. gas   23. matter   24. outermost energy levels
25. liquid   26. ion   27. electron cloud   28. quarks
29. alloys   30. periodic table

31. Adding heat causes the particles in a solid to move faster. When enough energy has been added, the individual particles in the solid move so rapidly that they break out of their fixed positions. This process results in the material becoming a liquid.

32. Atoms of the same element that differ from each other in mass number are called isotopes. Isotopes of any given atom have the same number of protons but a different number of neutrons.

33. A mixture is any material that contains two or more substances that are not chemically combined. The substances in a mixture keep their individual properties. A solution is a mixture in which one substance is uniformly dispersed in another substance.

34. Physical properties are characteristics such as melting point, hardness, color, etc., that can be observed without changing the composition of the substance. Chemical properties are characteristics that describe how a substance interacts with other substances to produce different kinds of matter.

35. The chemical formula for water is H_2O. This indicates that each water molecule consists of two atoms of hydrogen and one atom of oxygen. In a chemical formula, a subscript number is used after the symbol for an element to indicate how many atoms of that element are in a single molecule.

Guide to Tested Objectives

Objective 1: questions 1(A), 10(A), 11(A), 21(A), 23(A), 34(A); Objective 2: questions 3(A), 12(A), 15(A), 27(A), 28(B); Objective 3: questions 2(A), 13(A), 19(A), 20(B), 30(A); Objective 4: questions 14(A), 32(B); Objective 5: questions 4(A), 22(A), 25(A), 31(A); Objective 6: questions 5(A), 6(B), 7(A), 24(A), 26(A); Objective 7: questions 8(B), 9(A), 16(A); Objective 8: questions 17(B), 35(B); Objective 9: questions 18(A), 29(B), 33(A)

Answers to Chapter 9 Test

1. F	2. F	3. T	4. T	5. T	6. F	7. T	8. F	9. F	10. T
11. a	12. c	13. a	14. a	15. d	16. c	17. a	18. b	19. b	20. a

21. density = mass/volume   22. feldspar   23. streak   24. four
25. tetrahedral chains   26. radioactivity   27. refraction   28. conchoidal
29. native elements   30. rock-forming minerals

31. Quartz is a network silicate containing only silicon-oxygen tetrahedra; the bonds between the tetrahedra are very strong, making quartz extremely hard. Feldspar is also a network silicate. Unlike quartz, however, some tetrahedra in feldspar have atoms of metal instead of silicon. The bonds between these atoms are weaker than those between silicon and oxygen. Thus feldspar is not as hard as quartz.

32. First, scientists ask whether the substance is organic or inorganic; a mineral must be inorganic. Second, does the substance occur naturally in the earth; a mineral must be naturally occurring. Third, is the substance a solid; all minerals are solids. Fourth, does the substance have a definite chemical composition; minerals do.

33. Cleavage is the tendency of a mineral to split along certain flat surfaces. Cleavage is controlled by the types of bonds in the internal structure of the mineral.

34. The mining inspector is concerned primarily with the safety and environmental aspects of mining, while the mining engineer is concerned with the technical and economic aspects of mining.

35. The Mohs Scale is used to measure a mineral's hardness. The hardness of an unknown mineral is compared to the hardnesses of known minerals, which are grouped on a scale of relative hardness. The unknown mineral will scratch minerals softer than itself, and be scratched by minerals harder than itself.

Guide to Tested Objectives

Objective 1: questions 1(A), 9(A), 32(A), 34(A); Objective 2: questions 2(A), 3(B), 4(A), 11(A), 22(A), 30(A); Objective 3: questions 5(B), 6(A), 7(B), 29(A); Objective 4: questions 10(B), 14(A), 15(A), 24(A), 25(B), 31(B); Objective 5: questions 8(A), 12(B), 13(A), 16(B), 19(A), 20(A), 21(A), 23(A), 28(B), 33(A), 35(B); Objective 6: questions 17(A), 18(A), 26(B), 27(A)

M O D E R N   E A R T H   S C I E N C E

Answers to Chapter 10 Test

1. T	2. T	3. F	4. T	5. F	6. T	7. F	8. F	9. T	10. T
11. d	12. b	13. c	14. d	15. a	16. c	17. a	18. b	19. b	20. a

21. cools
22. compaction
23. fossils
24. concretions
25. formed
26. crystals
27. laccoliths
28. neck
29. breccia
30. cross-bedded

31. Foliation may occur in one of two ways. Pressure may flatten the mineral crystals in the original rock, squeezing them into parallel bands. Foliation may also occur as minerals of different densities separate into bands.

32. The size of mineral crystals in igneous rock depends largely on how quickly or slowly the magma cools. Extrusive rock forms when magma cools on the earth's surface. Since surface temperatures are relatively cool, the magma cools rapidly, which prevents the formation of large crystals.

33. Clastic sedimentary rock is formed through the cementing and compaction of rock fragments that have been deposited by water, wind, or ice.

34. Lava plateaus develop from lava that streams out of long cracks in the earth's surface. The lava then spreads out over a large area, filling in valleys and covering hills. The hardened layers of lava form a raised plateau.

35. As igneous rock is worn away and broken down by agents such as water and wind, rock fragments form. These fragments can then be cemented and compacted to form sedimentary rock.

Guide to Tested Objectives

Objective **1**: question 25(A); Objective **2**: questions 2(B), 7(A), 10(A), 35(B); Objective **3**: questions 1(A), 21(A), 26(A), 32(B); Objective **4**: question 11(B); Objective **5**: questions 3(B), 12(B), 16(A), 19(B), 20(B), 27(A), 28(A) 34(A); Objective **6**: questions 4(A), 5(A), 18(A), 22(A), 29(A), 33(A); Objective **7**: questions 6(A), 13(A), 15(A), 23(A), 24(A), 30(A); Objective **8**: questions 8(A), 17(A); Objective **9**: questions 9(A), 14(A) 31(A)

Answers to Chapter 11 Test

1. F	2. T	3. F	4. T	5. F	6. F	7. T	8. F	9. T	10. T
11. c	12. b	13. d	14. b	15. a	16. a	17. c	18. c	19. c	20. d

21. vein
22. hydrocarbons
23. magma
24. cap rock
25. petroleum
26. fusion
27. ocean
28. turbine
29. petrochemicals
30. uranium

31. One way to conserve minerals is to use more abundant materials, such as plastic, in place of minerals. Another way to conserve minerals is to recycle them, or use them over again. Metals such as iron, copper, and aluminum can be recycled.

32. During strip mining, topsoil and rocks that are displaced to expose the coal are left in steep slopes. When wet, rocks exposed during mining give off acids. Rain may carry the acids into nearby rivers and streams, causing harm to living things there.

33. Twice each day, water in the oceans moves toward and then away from the shore. These movements are called tides. High tide occurs when the water reaches its highest point along the shore, low tide when it reaches its lowest point. To make use of this tidal power, dams trap the water at high tide and release it at low tide. The water spins the turbines of electric generators to produce electricity.

34. Coal is formed when peat is subjected to pressure from overlying sediments. This pressure forces water out of the peat, leaving a material that is more dense and that contains a higher carbon content than the original peat. Because of the coal's higher density and carbon content, it burns longer and hotter than peat.

35. A nuclear fission reaction begins when U-235 atoms are struck by neutrons. The U-235 atoms split, releasing energy and more neutrons. These newly released neutrons strike more U-235 atoms, which release more energy and more neutrons to continue the chain reaction. The reaction can be slowed by lowering a neutron-absorbing material into the reactor. When the material is raised, neutrons become available again for the chain reaction.

Guide to Tested Objectives

Objective **1**: questions 3(A), 11(B), 14(B), 21(A), 23(A); Objective **2**: questions 6(A), 13(A), 16(A), 25(A), 29(A); Objective **3**: questions 5(A), 12(B), 15(B), 22(A), 34(B); Objective **4**: questions 7(A), 19(B), 20(B), 24(A); Objective **5**: question 10(A); Objective **6**: questions 1(A), 31(A); Objective **7**: questions 9(A), 32(B); Objective **8**: questions 4(A), 30(A), 35(B); Objective **9**: questions 26(A), 27(B); Objective **10**: question 17(A); Objective **11**: questions 2(A), 18(A), 28(A); Objective **12**: questions 8(A), 33(A)

M O D E R N   E A R T H   S C I E N C E

Answers to Unit 3 Test

1. T	2. T	3. F	4. T	5. F	6. F	7. F	8. T	9. F	10. T
11. F	12. T	13. T	14. F	15. b	16. c	17. c	18. a	19. d	20. b
21. d	22. a	23. d	24. a	25. b	26. c	27. b			

28. neutron

29. 21

30. solid

31. electrons

32. silicate and nonsilicate

33. conchoidal

34. 3.0 g/cm³

35. mineral

36. radioactive

37. sill

38. mafic

39. cools

40. stock

41. geode

42. hydrocarbons

43. ores

44. pressure

45. neutrons

46. petroleum

47. Physical properties are characteristics that can be observed without changing the composition of the substance. Color, hardness, and the ability to conduct heat and electricity are physical properties. Chemical properties are those characteristics that describe how a substance interacts with other substances to produce different kinds of matter.

48. Fluorescence is the ability of some minerals to absorb ultraviolet light and then emit visible light of various colors. Some minerals subjected to ultraviolet light will continue to glow after the ultraviolet light is turned off. Minerals that continue to glow have the property called phosphorescence.

49. During compaction, the weight of overlying sediment causes pressure, pushing the fragments together and squeezing out air and water from between the fragments. In cementation, water carries dissolved minerals through the sediments. These minerals are left between the fragments of sediment and provide a cement to hold the fragments together.

50. Carbonization occurs when partially decomposed plants are buried in swamp mud. Bacteria consume some of the plant material. These bacteria then release marsh gas, which includes methane (CH_4) and carbon dioxide (CO_2). As the gas escapes, the original compounds present in the plants gradually change, and only carbon remains.

Guide to Tested Objectives

Chapter 8—Objective 1: questions 18(A), 47(A); Objective 2: questions 1(A), 28(A); Objective 3: question 29(B); Objective 5: question 30(A); Objective 6: question 31(B); Objective 7: questions 15(A), 17(B); Objective 8: question 16(B); Objective 9: question 2(B). **Chapter 9**—Objective 1: questions 4(A), 32(A), 35(A); Objective 5: questions 3(A), 5(A), 6(A), 22(B), 23(A), 27(B), 33(B), 34(B); Objective 6: questions 36(A), 48(A). **Chapter 10**—Objective 1: question 19(A); Objective 2: question 9(A); Objective 3: questions 8(A), 20(A), 39(A); Objective 4: question 38(B); Objective 5: questions 10(A), 37(A), 40(B); Objective 6: question 49(A); Objective 7: questions 7(B), 41(A); Objective 8: question 21(B). **Chapter 11**—Objective 1: questions 24(A), 43(A); Objective 2: question 11(A); Objective 3: questions 26(B), 42(A), 50(B); Objective 4: 12(A), 44(A), 46(B); Objective 8: questions 13(B), 45(A); Objective 9: question 25(A); Objective 11: question 14(A).

M O D E R N   E A R T H   S C I E N C E

Answers to Chapter 12 Test

1. T	2. F	3. F	4. T	5. F	6. T	7. T	8. F	9. F	10. T
11. d	12. a	13. d	14. c	15. c	16. c	17. b	18. b	19. b	20. d

21. ice wedging   22. calcite   23. hydrolysis   24. laterite
25. erosion   26. sheet erosion   27. slump   28. monadnock
29. abrasion   30. mesas

31. Youthful mountains are rugged and have sharp peaks and deep, narrow valleys. Mature mountains have rounded peaks and gentle slopes.

32. A pedocal contains large amounts of calcium carbonate. They are less acidic than pedalfers and are very fertile. Pedocals form in areas that receive less than 65 cm of rain per year.

33. Rocks that are rich in feldspar form soils containing large amounts of clay; these soils consist of fine grains of silicate material containing aluminum and water. Rocks containing large amounts of quartz weather chemically to form sandy soils.

34. Contour plowing: the soil is plowed in circular bands that follow the contour of the land to reduce water erosion. Stripcropping: cover crops are planted next to crops that expose soil to erosion. Terracing: constructing steplike ridges that follow the contours of a sloped field reduces erosion. Crop rotation: helps to prevent gullying and sheet erosion.

35. Both soils form mainly as a result of mechanical weathering. This leads to thin soils consisting mainly of rock fragments called regolith.

Guide to Tested Objectives

Objective 1: questions 1(A), 6(B)-feature, 11(A), 14(B), 21(A), 29(B); Objective 2: questions 10(A), 23(B); Objective 3: questions 12(A), 13(A), 22(B); Objective 4: question 18(B); Objective 5: question 2(A); Objective 6: questions 4(A), 5(A), 7(A), 15(B), 19(A), 20(B), 33(B); Objective 7: questions 3(B), 17(A), 24(B), 32(B), 35(B); Objective 8: questions 16(A), 25(A); Objective 9: questions 8(A), 9(B), 27(A); Objective 10: questions 26(A), 34(A); Objective 11: questions 28(A), 30(A), 31(B)

Answers to Chapter 13 Test

1. T	2. T	3. F	4. F	5. F	6. T	7. T	8. T	9. T	10. T
11. c	12. b	13. a	14. d	15. a	16. b	17. c	18. a	19. d	20. b

21. local water budget   22. cools   23. desalination   24. a gully
25. natural levees   26. saltation   27. wind gap   28. channel
29. a floodplain   30. a floodway

31. Factors that affect the local water budget include temperature, presence of vegetation, wind, and the amount and duration of precipitation.

32. The process by which liquid water changes into water vapor is called evaporation. Water vapor also enters the air by transpiration, a process in which plants give off water vapor into the atmosphere.

33. Near the headwaters, the gradient is steep and the stream generally has a high velocity. Near the mouth of the stream, the gradient often becomes less steep and velocity decreases.

34. The total load a stream can carry is greatest when a large volume of water is flowing swiftly. When the velocity of the water decreases, the ability of the stream to carry its load generally also decreases. As a result, part of the stream load is deposited as sediment. When velocity increases again, some or all of the sediment is carried away.

35. Stream deposition is the process by which a stream entering a larger body of water deposits its load. As the Mississippi River enters the Gulf of Mexico, it deposits its load and slowly adds on to the coastline of the United States.

Guide to Tested Objectives

Objective 1: questions 1(B), 19(B), 22(A), 32(A); Objective 2: questions 16(A), 18(A), 21(A), 31(A); Objective 3: questions 7(A), 23(A); Objective 4: questions 3(A), 4(A), 8(A), 24(A), 28(A); Objective 5: questions 5(A), 11(A), 14(A), 26(A), 27(A), 33(B); Objective 6: questions 15(B), 17(A), 20(A); Objective 7: questions 2(B)-feature, 6(A), 13(A), 34(B); Objective 8: questions 9(A), 25(A), 29(A), 35(B); Objective 9: questions 10(A), 12(A), 30(A)

M O D E R N   E A R T H   S C I E N C E

Answers to Chapter 14 Test

1. T	2. T	3. T	4. F	5. F	6. T	7. F	8. T	9. T	10. F
11. b	12. a	13. b	14. c	15. a	16. a	17. d	18. d	19. d	20. c

21. water table 22. stalactites 23. porosity 24. groundwater
25. geysers 26. ordinary springs 27. impermeable 28. capillary action
29. gravity 30. sinkholes

31. Mud pots form when the rock surrounding a hot spring is chemically weathered by volcanic gases dissolved in the water. The weathered rock mixes with hot water to form a sticky, liquid clay that bubbles at the surface.
32. Groundwater may become polluted by waste dumps, leaking underground storage tanks for toxic chemicals, agricultural and lawn fertilizers and pesticides, and leaking sewage systems. If too much groundwater is pumped from an aquifer near the ocean, the water table may drop below sea level, allowing salt water to flow in and contaminate the aquifer.
33. Gradient is the steepness of any slope, including that of a water table. The velocity of water flow increases as its gradient increases, but slows as its gradient decreases.
34. Natural bridges form when the roof of a cavern collapses in several places, leaving stretches of uncollapsed rock between the sinkholes. A natural bridge can also form when a surface river enters a crack in a rock formation, runs underground, and then reemerges. The river weathers the rock it passes through, and eventually forms a natural bridge.
35. The rock layer has high permeability and low porosity. The porosity of rock is influenced by the size and arrangement of the particles forming the rock. If the particles are many different sizes, small particles fill in the spaces between larger ones, making the rock less porous. If the particles are packed tightly together, there are few open spaces to hold water.

Guide to Tested Objectives

Objective 1: questions 3(A), 16(B), 23(A), 27(A), 35(B); Objective 2: questions 13(A), 15(A), 21(A), 24(A), 28(B); Objective 3: questions 1(A), 9(A), 26(A), 29(A), 33(B); Objective 4: questions 17(A), 18(B), 32(A); Objective 5: questions 6(A), 8(B), 19(B), 20(B); Objective 6: questions 5(A), 25(A), 31(A); Objective 7: questions 2(A), 7(A), 10(A), 12(A), 22(A), 30(A)-feature; Objective 8: questions 4(A), 11(B), 14(A), 34(A)

Answers to Chapter 15 Test

1. F	2. T	3. F	4. T	5. F	6. T	7. F	8. T	9. T	10. F
11. a	12. c	13. b	14. d	15. a	16. c	17. d	18. a	19. d	20. d

21. crevasses 22. oxygen 23. ice age 24. snowfield
25. valley glacier 26. Agassiz 27. internal plastic flow
29. outwash plains 30. minerals

31. Snow accumulates in snowfields year after year. The partial melting and refreezing of the snow crystals changes them into grainy ice called firn. In the deepest layers of accumulated snow, the pressure of the overlying layers becomes so great that the air between the ice grains is squeezed out. Each year more snow is added on top of the ice layers. Eventually, the weight of the ice becomes great enough to cause the body of ice to move, and a glacier is formed.
32. The weight of the ice in a glacier exerts enough pressure to melt the ice where it comes in contact with the ground. The water from the melted ice acts as a lubricant that allows the base of the ice to slide over the underlying rock.
33. Large lake basins fill with rainfall and meltwater when the glaciers retreat. In lakes that have no outlet streams, water can only leave by evaporation. As water evaporates, salt that is dissolved in the water is left behind and the water becomes increasingly salty. This results in the formation of salt lakes.
34. One change is in the shape of the earth's orbit, which changes back and forth between nearly circular and more elongated. A second change is in the degree of tilt of the earth's axis. A third periodic change results from the precession, or circular motion, of the earth's axis.
35. An increase in solar energy would likely cause the continental ice sheets of the world, such as those in Greenland and Antarctica, to begin melting. This increase in water in the oceans would significantly raise the sea level. Coastlines would then be flooded for some distance inland, depending on the flatness of the coastal region.

Guide to Tested Objectives

Objective 1: questions 3(A), 11(A), 12(A), 16(B), 24(A), 31(A); Objective 2: questions 13(A), 25(A); Objective 3: questions 1(A), 2(A), 6(B)-feature, 15(A), 21(A), 27(A), 32(A); Objective 4: questions 5(A), 7(A), 8(A), 14(A), 28(A); Objective 5: questions 9(A), 18(A), 19(A), 20(B), 29(A); Objective 6: questions 4(A), 26(A)-feature, 30(A), 33(A); Objective 7: questions 23(A), 35(B); Objective 8: questions 10(A), 17(A), 22(B), 34(B)

M O D E R N   E A R T H   S C I E N C E

Answers to Chapter 16 Test

1. T	2. F	3. F	4. T	5. F	6. T	7. T	8. F	9. T	10. T
11. b	12. c	13. c	14. c	15. d	16. a	17. b	18. a	19. d	20. a

21. saltation   22. deflation   23. deflation hollow   24. loess
25. wave-cut terrace   26. berms   27. headland   28. lagoons
29. parabolic dune   30. sea cave

31. Eroded material may be deposited some distance from the shore, creating an extension to the wave-cut terrace called a wave-built terrace. The water above the terrace is shallow, and waves lose much of their energy passing through this shallow water. As the energy of a wave lessens over a terrace, the rate of erosion of the cliff is greatly reduced.

32. The waves force salt water and air into small cracks in the rock. Substances in the air and water produce a chemical action that may enlarge the cracks. The enlarged cracks, in turn, provide an increased surface area for physical and chemical weathering.

33. These floods have been caused by three factors—erosion, sinking land, and rising sea level. Erosion of barrier beaches allows the high tides to reach further into Venice. The weight of added sediments is causing the city to sink. The withdrawal of too much groundwater as well as the weight of the buildings within the city has caused additional sinkage. Also, like oceans worldwide, the nearby Adriatic Sea is rising.

34. As waves wash up on the shore, they move sand and small rock fragments forward. In retreating, the water moves some fragments away from the shore. Beaches form where the amount of rock fragments moving toward the shore is greater than the amount moving away from the shore.

35. Black sand mixed with shell and coral fragments would most likely compose such a beach. Black sand comes from the volcanic rock that is commonly found on Hawaii and other volcanic islands. A beach located near an offshore coral reef would also consist of fragments of shells and coral that are washed ashore.

Guide to Tested Objectives

Objective 1: questions 1(A), 11(A), 12(A), 21(A), 22(A), 23(A); Objective 2: questions 2(A), 13(A), 24(A), 29(B); Objective 3: questions 3(A), 14(B), 15(A), 25(A), 30(A), 31(B), 32(B); Objective 4: questions 16(A), 26(A), 34(A), 35(B); Objective 5: questions 5(A), 6(B), 27(A); Objective 6: questions 7(B), 8(A), 9(A), 10(A), 17(A), 33(B)-feature; Objective 7: questions 4(A), 28(A); Objective 8: questions 19(B), 20(B); Objective 9: question 18(A)

Answers to Unit 4 Test

1. F	2. T	3. F	4. F	5. T	6. T	7. T	8. F	9. T	10. F
11. F	12. T	13. T	14. F	15. T	16. c	17. d	18. b	19. c	20. a
21. d	22. c	23. d	24. c	25. c	26. a	27. a	28. c	29. b	

30. snowfall   31. zone of saturation   32. erosion   33. glacial drift
34. geyser   35. A horizon   36. quartz   37. natural levees
38. the windward side   39. suspended load   40. tombolos   41. the mature stage
42. glacial melting   43. discharge   44. barrier island 45. groundwater
46. water table

47. As the walls and floor of a valley are scraped away by a glacier, the original stream-cut V-shape of the valley is changed into a glacial U-shape. Because glacial action is the only means by which a valley can acquire a U-shape, scientists can easily tell whether a valley has been glaciated.

48. Plants and animals are important agents of mechanical weathering. The roots of plants can work their way into cracks in rocks, creating pressure that wedges the rock apart. Some plants also produce acids that chemically weather rocks. The digging activities of burrowing animals contribute to weathering by exposing new rock surfaces to weathering agents.

49. First, a coral reef forms around a tropical volcanic island. Then, as the ocean lithosphere bends under the weight of the volcano, both the volcano and the reef sink. The coral reef builds higher because the animals can only live near the surface. When the volcano is no longer active, the island may sink below the surface. The reef, however keeps growing upward, and eventually only a nearly circular reef remains above the surface, surrounding a shallow lagoon.

50. Factors that affect a local water budget include temperature, the presence of vegetation, wind, and the amount and duration of rainfall. When precipitation exceeds evapotranspiration and runoff, the result is moist soil and possible flooding. When evapotranspiration exceeds precipitation, soil becomes dry. Vegetation reduces runoff. Winds increase the rate of evapotranspiration.

Guide to Tested Objectives

Chapter 12—Objective 1: question 48(A); Objective 2: questions 5(B)-feature, 18(A); Objective 5: question 4(A); Objective 6: question 20(A); Objective 7: question 35(B); Objective 8: question 32(A); Objective 9: question 21(A); Objective 11: question 17(A). Chapter 13—Objective 1: question 24(A); Objective 2: questions 8(B), 50(B); Objective 4: question 43(A); Objective 5: questions 9(A), 39(A); Objective 6: questions 25(A), 34(A), 41(A); Objective 7: questions 10(A), 37(A).
Chapter 14—Objective 1: question 13(A); Objective 2: questions 31(A), 45(B), 46(B); Objective 5: question 15(B); Objective 6: question 14(A); Objective 7: questions 27(A), 28(A). Chapter 15—Objective 3: questions 2(B), 12(A); Objective 4: questions 1(A), 19(A), 26(A), 47(B); Objective 5: questions 22(B), 33(A); Objective 7: questions 23(A), 30(A); Objective 8: question 3(A). Chapter 16—Objective 1: question 36(A); Objective 2: questions 29(A), 38(A); Objective 3: question 6(B); Objective 4: question 44(A); Objective 5: questions 7(A), 40(A); Objective 6: question 42(B); Objective 7: question 16(A); Objective 8: question 49(B); Objective 9: question 11(A).

HRW material copyrighted under notice appearing earlier in this work.

207

M O D E R N   E A R T H   S C I E N C E

Answers to Chapter 17 Test

1. T	2. T	3. F	4. F	5. T	6. F	7. T	8. T	9. T	10. T
11. b	12. a	13. a	14. c	15. b	16. d	17. d	18. c	19. d	20. d

21. uniformitarianism
22. half-life
23. nonconformity
24. daughter element
25. paleontologists
26. sedimentation
27. casts
28. gastroliths
29. index fossils
30. evolution

31. The fossil must be present in rocks scattered over a wide area of the earth's surface, and must be clearly distinguishable from other fossils. The organism from which the index fossil formed must have lived for only a short span of geologic time. Finally, index fossils must occur in fairly large numbers within rock layers.

32. If scientists can measure the rate at which a stream erodes its bed, they can determine the average age of the stream. This method of dating is not dependable for older features because the rates of erosion have varied greatly throughout history.

33. Mineral solutions such as groundwater remove the original organic matter in an organism and replace it with mineral matter. Some common petrifying minerals are silica, calcite, and pyrite.

34. Radioactive decay is the process by which an element spontaneously emits protons, electrons, neutrons, and energy. The radioactive element, called the parent element, decays until it forms a stable, nonradioactive daughter element.

35. According to the law of superposition, all rocks beneath an unconformity are older than those rocks above the unconformity. If a fault or intrusion cuts through rock layers or other rock features, the fault or intrusion is younger than all the rocks and rock features it cuts through.

Guide to Tested Objectives

Objective 1: questions 6(A), 21(A); Objective 2: questions 2(A), 15(A), 18(B), 19(A); Objective 3: questions 14(B), 16(A), 23(B); Objective 4: questions 17(A), 35(B); Objective 5: questions 3(A), 32(B); Objective 6: questions 4(A), 12(A); Objective 7: questions 1(A), 5(A), 20(A), 22(A), 24(A), 34(A); Objective 8: questions 9(A), 11(A), 13(A), 25(A), 26(A), 33(B); Objective 9: questions 7(A), 8(A), 10(A), 27(A), 28(A); Objective 10: questions 29(A), 30(A), 31(B)

Answers to Chapter 18 Test

1. T	2. F	3. T	4. T	5. T	6. F	7. F	8. T	9. T	10. F
11. b	12. d	13. c	14. d	15. a	16. c	17. d	18. d	19. b	20. c

21. Ordovician
22. radioactive dating
23. Charles Darwin
24. Tertiary and Quaternary
25. Mesozoic Era
26. Pleistocene
27. trilobites
28. water
29. meteorite impact
30. Cretaceous

31. The fossil record of the Cenozoic Era is very abundant. Smaller time divisions are largely based on changes in the fossil content of rock layers.

32. When environmental changes occur, some organisms can adapt to live with the changes better than others. The organisms that adapt compete more successfully in the new environment and survive to reproduce, while the organisms that cannot adapt become extinct.

33. These rocks are very old and have had lots of time to be deformed and altered by crustal activity. Therefore, original rock layers and their order are rarely identifiable.

34. Carboniferous means carbon-bearing. It was named for the wealth of coal- and oil-bearing deposits of that age.

35. The Mesozoic was warm and humid. Reptiles are cold-blooded and require external warmth to maintain their body functions at levels sufficient to pursue normal, survival-related activities.

Guide to Tested Objectives

Objective 1: questions 11(A), 14(A), 22(B), 28(A), 29(B)-feature, 32(B); Objective 2: questions 2(A), 3(A), 12(A), 13(A); Objective 3: questions 1(A), 10(A), 15(A), 16(A), 23(A), 27(A), 33(A); Objective 4: questions 4(A), 5(A), 9(A), 17(B), 21(B), 34(A); Objective 5: questions 6(A), 7(A), 25(A) 30(B), 35(B); Objective 6: questions 8(A), 18(B), 19(B), 20(B), 24(B), 26(A), 31(A)

M O D E R N   E A R T H   S C I E N C E

Answers to Chapter 19 Test

1. F	2. T	3. F	4. F	5. T	6. T	7. T	8. T	9. T	10. F
11. b	12. d	13. c	14. d	15. a	16. b	17. a	18. a	19. a	20. c

21. Cenozoic Era   22. outcrop   23. the law of superposition   24. the South Atlantic Ocean
25. radioactive dating   26. sand dunes   27. Mesozoic Era   28. bacteria
29. Cenozoic   30. Colorado Plateau

31. In North America, Mexico's Baja Peninsula and the portion of California west of the San Andreas fault will have moved to where Alaska is today. If this plate movement occurs as predicted, Los Angeles will one day be located north of where San Francisco is today.
32. Twenty thousand years ago, a land bridge existed between Siberia and Alaska across the Bering Strait. Ice sheets that covered much of North America held huge quantities of water. As a result, the sea level was much lower than it is today. The first Americans were probably hunters who crossed the bridge following herds of bison and mammoth.
33. If the Grand Canyon area were uplifted again, the Colorado River would likely cut deeper into the Colorado Plateau, exposing deeper and deeper rock layers. As a result of this increased erosion, the Grand Canyon would become deeper.
34. Paleontologists can trace changes over time in many species by examining the fossils found in successive rock layers. Analysis of the fossils found in older, lower layers through younger, upper layers suggests that early life-forms gradually became extinct or changed into different forms.
35. The Grand Canyon probably began as a small valley cut by the Colorado River. Slowly, the area that included the valley was lifted to form a higher plateau called the Colorado Plateau. As this plateau was uplifted, the river cut down through the rock layers. Subsequently, weathering and mass movement has helped widen the canyon.

Guide to Tested Objectives

Objective 1: questions 2(A), 3(A), 4(A), 5(A), 19(A), 20(A); Objective 2: questions 1(A), 6(A), 7(A), 13(A), 24(A); Objective 3: questions 8(A), 9(A), 11(A), 12(B), 14(A), 15(A), 28(B); Objective 4: questions 10(A), 16(B), 21(A), 27(B), 29(B), 30(A), 31(B), 32(A); Objective 5: questions 17(B), 18(B), 22(A), 23(A), 26(A), 33(B), 35(A); Objective 6: questions 25(A), 34(A)

Answers to Unit 5 Test

1. F	2. T	3. T	4. F	5. F	6. T	7. T	8. T	9. T	10. T
11. T	12. F	13. T	14. F	15. T	16. F	17. a	18. d	19. d	20. c
21. c	22. d	23. a	24. b	25. d	26. d	27. c	28. c	29. b	30. a

31. Paleozoic   32. at the top   33. **Gondwanaland**   34. trilobites
35. amber   36. Laurasia   37. Canadian Shield   38. theory of evolution
39. Panthalassa   40. period   41. coprolites   42. **trace fossil**
43. petrification   44. extinction   45. Precambrian

46. Sedimentary layers tend to form with the coarsest-grained rock at the bottom. Another clue to the original position of the rock is found in the bedding plane. Cross-beds are curved at the bottom and cut off at the top by erosion. Ripple marks can also be helpful in determining the order of rock layers. The peaks of ripple marks point upward.
47. Northward drift brought the landmass into warmer climates. Ice covering the continent melted, raising the level of the oceans. This triggered global warming. The northward drift also resulted in the increase of the number of species of land plants and animals.
48. The early humans were hunters. Their successful hunting may have contributed to the extinction of some animals.
49. The supercontinent Pangaea broke up in the Mesozoic Era. The new, smaller continents drifted and collided, uplifting these mountains.
50. Carbon-14 and carbon-12 occur naturally in the atmosphere. All living organisms take in C-12 and C-14 in known ratios. When an organism dies, that ratio changes as the radioactive C-14 decays to nitrogen-14, but the stable C-12 remains constant. To determine the age of a sample of organic material, scientists find the ratio of C-12 to C-14, and then use the half-life of C-14 (about 5,730 years) to determine the age of the sample.

Guide to Tested Objectives

Chapter 17—Objective 1: questions 29(A), 30(A); Objective 2: question 46(B); Objective 3: questions 6(A), 15(B), 23(A), 38(A); Objective 4: questions 14(A), 18(A); Objective 7: questions 7(B), 21(B), 50(B); Objective 8: questions 22(A), 35(A), 43(A); Objective 9: questions 41(A), 42(A); Objective 10: question 1(A). **Chapter 18**—Objective 1: questions 3(A), 32(A); Objective 2: questions 11(A), 40(A); Objective 3: 12(A), 44(A), 45(B); Objective 4: questions 20(A), 34(B); Objective 5: questions 16(A), 28(B), 49(B); Objective 6: questions 8(A), 19(B), 48(B). **Chapter 19**—Objective 1: questions 9(B), 31(B), 33(A), 39(A), 47(B); Objective 2: questions 2(A), 13(A), 36(A); Objective 3: question 37(A); Objective 4: questions 17(A), 25(A)-feature, 26(A)-feature; Objective 5: questions 4(A), 24(B), 27(B); Objective 6: questions 5(A), 10(A).

M O D E R N   E A R T H   S C I E N C E

Answers to Chapter 20 Test

1. T	2. F	3. F	4. F	5. T	6. T	7. F	8. F	9. T	10. T
11. c	12. a	13. b	14. c	15. a	16. b	17. b	18. d	19. b	20. c

21. continental rise   22. sea   23. clay   24. seamounts
25. Grand Banks   26. submarine canyons   27. abyssal plains   28. landslides
29. nodules   30. oceanography

31. During glacial periods, the glaciers hold great amounts of water and sea level falls. Later, as the glaciers melt and water is added to the ocean, sea level rises again.

32. Seamounts may rise above the ocean surface and be eroded to flattened tops by waves. As the action of plate tectonics carries seamounts away from the area in which they were formed, the ocean crust sinks and pulls the seamounts beneath the ocean surface. These flat-topped, submerged seamounts are called guyots.

33. Glaciers pick up great quantities of rock as they move across land. When an iceberg breaks off the glacier and melts, the rock material sinks to the ocean floor.

34. A sonar system consists of a transmitter and a receiver. The transmitter sends out a continuous series of sound waves from a ship to the ocean floor. The sound waves, traveling at the speed of about 1,500 m/sec, bounce off the ocean floor and are reflected back up to the receiver. The time taken by the waves to complete the trip is used to determine the depth of the water.

35. With a narrow continental shelf, the currents would probably be less strong. This decrease would affect the food supply for fish, thereby decreasing the number of fish present. In deep waters, there would exist no available shellfish. They would be restricted to the small area of continental shelf and, therefore, be fewer in number.

Guide to Tested Objectives

Objective **1**: questions 4(A), 5(A), 11(B), 19(B), 22(A); Objective **2**: questions 1(A), 7(A), 16(B), 30(A), 34(B); Objective **3**: questions 8(A), 10(A), 12(A), 17(A), 21(A), 25(B)-feature, 26(A), 28(A), 31(A), 35(B)-feature; Objective **4**: questions 2(A), 18(A), 24(A), 27(A), 32(A); Objective **5**: questions 9(A), 13(A), 14(A), 15(A), 29(A), 33(B); Objective **6**: questions 3(A), 6(B), 20(A), 23(A)

Answers to Chapter 21 Test

1. T	2. T	3. F	4. T	5. F	6. T	7. F	8. T	9. T	10. T
11. d	12. d	13. b	14. c	15. a	16. b	17. b	18. b	19. a	20. d

21. salinity   22. blue   23. upwelling   24. phytoplankton
25. intertidal   26. nekton   27. pelagic   28. desalination
29. temperature   30. aquaculture

31. The thermocline exists because the water near the surface becomes less dense as it is warmed by heat from the sun. This warm water cannot mix easily with the cold, dense water below. Thus, a thermocline marks the separation between warm surface water and the colder deep water.

32. Lead can cause problems in ocean food webs. As animals eat other animals, the amount of lead builds up in the bodies of the predators. In some areas, the lead concentration in fish has made them inedible.

33. Distillation involves heating ocean water to remove the salt. Heat causes the liquid water to evaporate, leaving dissolved salts behind. When the water vapor condenses, the result is pure, fresh water.

34. Evaporation increases the salinity of ocean water. During evaporation, only water molecules are removed. Dissolved salts and other solids remain in the ocean. When the rate of evaporation is high, the relative amount of dissolved solids in surface water increases.

35. As the sea level dropped, the zones would have to move farther toward the open ocean than they currently are. Sunlight would most likely penetrate more of the shallower ocean, and plants would grow in the expanded neritic zone. The three deepest bottom zones, the bathyal, abyssal, and hadal zones, would probably occupy a smaller area because the oceans would be generally more shallow.

Guide to Tested Objectives

Objective **1**: questions 11(A), 12(B), 13(B), 17(A), 21(A), 34(A); Objective **2**: questions 2(A), 3(A), 4(A), 22(A), 29(A), 31(A); Objective **3**: questions 1(A), 9(A), 15(B), 23(A); Objective **4**: questions 5(A), 6(A), 24(A), 26(A); Objective **5**: questions 7(A), 19(A), 20(A), 25(A), 27(A), 35(B); Objective **6**: questions 8(B)-feature, 14(A), 16(B), 18(B), 28(A), 30(A), 33(A); Objective **7**: questions 10(A), 32(B)

M O D E R N   E A R T H   S C I E N C E

Answers to Chapter 22 Test

1. T	2. T	3. T	4. T	5. F	6. T	7. F	8. F	9. F	10. T
11. c	12. d	13. c	14. a	15. a	16. d	17. b	18. b	19. a	20. b

21. surface current   22. Bay of Fundy   23. wave period   24. refraction
25. tsunami   26. tides   27. the moon   28. a whitecap
29. tidal range   30. fetch

31. As a wave moves into shallow water, the bottom of the wave is slowed by friction. The top of the wave continues to move at its original speed and gets farther and farther ahead of the bottom of the wave. The top of the wave finally topples over and forms a breaker, a foamy mass of water that washes onto the shore.

32. When water is cooled, it contracts and the water molecules move closer together. This contraction makes the water more dense and, as a result, it sinks. When water is warmed, it expands and the water molecules move father apart. Warm water is thus less dense and remains above cold water.

33. The height of a tsunami is low in the open ocean because the entire depth of the water is involved in the wave motion. Near the shore, the height of the tsunami greatly increases as its speed decreases.

34. The tidal bulge that occurs on the side of the earth opposite the moon is caused by the mutual revolution of the earth and moon around a common center of gravity. Their spinning around each other generates outward forces that causes the ocean to bulge from the earth's surface.

35. A period of spring tides is a better time to plan the site of an oceanside building. The highest tide of the month occurs during this period and will indicate the maximum distance the water will travel up the shore.

Guide to Tested Objectives

Objective 1: questions 1(A), 2(A), 3(A), 4(A), 9(A), 11(B), 21(A); Objective 2: questions 10(A), 14(A), 17(A), 32(A); Objective 3: questions 12(A), 18(B), 19(A), 20(A), 23(A), 28(A), 30(B); Objective 4: questions 5(A), 13(A), 24(A), 25(A), 31(A), 33(A); Objective 5: questions 6(A), 15(A), 26(A), 27(A), 34(B), 35(B); Objective 6: questions 7(A), 8(A), 16(B), 22(B)-feature, 29(A)

Answers to Unit 6 Test

1. T	2. F	3. T	4. T	5. F	6. F	7. T	8. F	9. T	10. T
11. F	12. F	13. T	14. T	15. F	16. F	17. c	18. c	19. a	20. a
21. b	22. a	23. a	24. b	25. b	26. d	27. b	28. d	29. c	30. a
31. a									

32. neritic zone   33. trade winds   34. breaker   35. bathyscaph
36. siliceous ooze   37. Pacific Ocean   38. phytoplankton   39. spring tide
40. turbidity current   41. calcium carbonate   42. abyssal zone   43. distillation
44. density

45. During the glacial periods, when continental ice sheets held great amounts of water, sea level fell. Due to this low sea level, parts of the continental shelves were exposed to weathering and erosion.

46. The particles move in circular paths. During a single wave period, each water particle moves in one complete circle. At the end of the wave period, a circling water particle ends up almost exactly where it started.

47. When ocean water freezes, the first ice crystals that form are free of salt. The salt remains in pockets of liquid water in the ice. The ice can then be removed and melted to obtain fresh water.

48. Because the sun cannot directly heat ocean water below its uppermost regions, the temperature of the water drops sharply as the depth increases. In most places in the ocean, this sudden temperature drop begins not far below the surface. This zone of rapid temperature change is called the thermocline. The warm uppermost water cannot mix easily with the cold, dense water below; a thermocline marks the distinct separation between the warm surface water and the colder deep water.

49. Along the Atlantic Coast, tides follow a semidiurnal, or twice-daily, pattern. There are two high tides and two low tides each day, with a fairly regular tidal range. Along the Pacific Coast, tides follow a mixed pattern, with an irregular tidal range.

50. Much of a meteorite vaporizes as it enters the earth's atmosphere. What remains falls to the earth's surface. Because so much of the earth's surface is ocean, most meteorite fragments fall into the ocean and become part of the sediments on the ocean floor.

Guide to Tested Objectives

Chapter 20—Objective 1: questions 2(A), 19(A), 37(A); Objective 2: question 35(B); Objective 3: questions 15(A), 20(A), 45(A); Objective 4: questions 1(A), 10(A), 18(A); Objective 5: questions 3(A), 16(A), 24(B), 27(A), 50(A); Objective 6: questions 36(A), 41(A). Chapter 21—Objective 1: questions 5(A), 8(B), 21(A); Objective 2: questions 9(A), 44(B), 48(B); Objective 3: question 12(A); Objective 4: questions 29(A), 38(A); Objective 5: questions 30(A), 32(A), 42(B); Objective 6: questions 4(A), 31(A), 43(A), 47(A). Chapter 22—Objective 1: questions 14(A), 33(A); Objective 2: questions 6(A), 40(A); Objective 3: questions 17(A), 25(B), 46(A); Objective 4: questions 23(A), 28(B), 34(B); Objective 5: questions 13(A), 26(A), 39(A); Objective 6: questions 7(A)-feature, 11(A), 22(A), 49(A).

M O D E R N   E A R T H   S C I E N C E

Answers to Chapter 23 Test

1. F	2. T	3. T	4. T	5. F	6. F	7. T	8. F	9. T	10. T
11. c	12. a	13. b	14. a	15. b	16. d	17. c	18. b	19. b	20. d

21. fossil fuels   22. breezes   23. Coriolis effect   24. exosphere
25. infrared rays   26. climate   27. a temperature inversion   28. jet streams
29. millibars   30. conduction

31. Because air is such a poor conductor of heat, it takes several hours before enough heat has been absorbed and reradiated from the ground to warm the air. The sun is highest at noon, but there is a time lag during which the air is heating.

32. Ozone absorbs ultraviolet radiation from the sun. A reduction in ozone would result in more of this harmful radiation striking the earth.

33. On a few occasions, severe air pollution has caused a large number of deaths. Also, studies have shown that long-term exposure to pollution reduces a person's ability to resist many illnesses.

34. During the day when the air is warmer above the land, the warm air rises, and the cool air from above the water moves in to replace it. This is a sea breeze. When the land becomes cooler than the water at night, the breeze shifts direction and becomes a land breeze.

35. The source of the moon's light is the sun. The light is reflected by the moon's surface toward the earth.

Guide to Tested Objectives

Objective 1: questions 11(A), 16(A), 17(A), 26(A), 32(A); Objective 2: questions 2(A), 8(A), 10(A), 29(A); Objective 3: questions 5(A), 13(A), 14(B), 24(A); Objective 4: questions 4(A), 21(A), 27(A), 33(A); Objective 5: questions 7(A), 15(A), 35(B); Objective 6: questions 3(A), 25(A), 31(A); Objective 7: questions 6(B), 12(A), 30(B); Objective 8: questions 18(B), 19(B), 20(A), 23(A), 28(A); Objective 9: questions 1(A), 9(A)-feature, 22(A), 34(B)

Answers to Chapter 24 Test

1. T	2. F	3. T	4. T	5. F	6. F	7. F	8. T	9. T	10. F
11. b	12. c	13. c	14. d	15. b	16. d	17. b	18. a	19. c	20. d

21. latent heat   22. humidity   23. conduction   24. condensation nuclei
25. cumulus clouds   26. ground fog   27. temperature   28. silver iodide
29. cloud seeding   30. water content

31. In ice, water molecules are held almost stationary in a definite crystalline arrangement. As heat is added, the molecules begin to move more rapidly and the ice melts. In liquid water, the molecules move around each other but they stay closer together. As water is heated, molecules move even faster and collide. Collisions cause molecules to move rapidly enough to evaporate and escape to become water vapor.

32. A psychrometer consists of a dry-bulb thermometer and a wet-bulb thermometer. The two thermometers are whirled to circulate air around the bulbs. As water on the wet-bulb thermometer evaporates, heat is withdrawn from the thermometer. The temperature drop seen in the wet-bulb thermometer depends on the humidity of the air. A table is used to translate the temperature difference into relative humidity.

33. In supercooled clouds, most of the water exists as supercooled water droplets. There are few freezing nuclei. Precipitation forms as water molecules evaporate from supercooled water droplets and condense into the ice crystals. The ice crystals increase in size until they are large enough to fall as snow and rain.

34. Hailstones often consist of alternating clear and cloudy layers of ice. The clear ice forms as a hailstone passes through a layer of very moist air. The cloudy layer forms when water droplets with air bubbles between them freeze on the surface.

35. Water from the warmer air surrounding the glass cools to its dew point and condenses on the cooled surface of the glass. This is also how dew forms on many surfaces overnight.

Guide to Tested Objectives

Objective 1: questions 1(A), 11(A), 21(B), 31(A); Objective 2: questions 2(A), 12(B), 22(A), 32(B); Objective 3: questions 3(A), 13(A), 23(A), 35(B); Objective 4: questions 4(A), 14(A), 24(A); Objective 5: questions 5(A), 19(A), 20(B), 25(A); Objective 6: questions 6(A), 15(A), 26(A); Objective 7: questions 7(A)-feature, 16(A), 27(A), 34(A); Objective 8: questions 8(A), 17(B), 28(A), 33(B); Objective 9: questions 9(A), 29(A); Objective 10: questions 10(A), 18(A), 30(A)

M O D E R N   E A R T H   S C I E N C E

Answers to Chapter 25 Test

1. F   2. T   3. T   4. F   5. T   6. F   7. F   8. F   9. T   10. F
11. a   12. c   13. b   14. a   15. b   16. d   17. a   18. d   19. c   20. d
21. maritime polar Pacific   22. stationary   23. counterclockwise   24. 180°
25. maritime   26. 5 days   27. lightning   28. radiosonde
29. satellites   30. occluded

31. A thunderstorm meets high-altitude horizontal winds. These winds cause the rising air to rotate. A narrow, funnel-shaped, rapidly spinning extension may develop below one of the storm clouds. This extension reaches downward and may or may not touch the ground.

32. Cold, dense polar air creates high-pressure zones at the poles. Warmer, less dense air creates low-pressure zones at the equator. As cold polar air sinks, it pushes toward the equator along the earth's surface. Warm equatorial air rises and moves toward the poles, where it cools and sinks. These conditions create three convection cells in both the Northern and Southern Hemispheres.

33. The three stages of a thunderstorm are the cumulus stage, the mature stage, and the final stage. These storms occur when warm, moist air rises and condenses. As the air cools, it swells upward, forming an anvil-shaped cumulonimbus cloud. Electrical discharges in the cloud cause thunder and lightning.

34. A liquid thermometer consists of a liquid sealed in a glass tube. A rise in temperature causes the liquid to expand and move up the tube. A drop in temperature causes the liquid to contract and move down the tube. A scale marked on the glass tube indicates the temperature by indicating the amount of expansion or contraction of the liquid.

35. Releasing large quantities of ions near the ground can modify the electrical properties of small cumulus clouds. Seeding the clouds with silver iodide also seems to modify lightning patterns.

Guide to Tested Objectives

Objective 1: questions 1(A), 32(B); Objective 2: questions 2(A), 11(B), 12(A), 21(A), 25(A); Objective 3: questions 3(A), 13(A), 22(A), 30(A); Objective 4: questions 4(A), 18(B), 23(A); Objective 5: questions 5(A), 14(A), 31(A), 33(B); Objective 6: questions 6(A), 16(A), 24(A), 34(A); Objective 7: questions 7(A), 28(A), 29(A); Objective 8: questions 8(A), 17(B), 19(B), 20(B); Objective 9: questions 9(A), 10(A), 15(A), 26(B), 27(A), 35(B)

Answers to Chapter 26 Test

1. F   2. T   3. T   4. F   5. T   6. T   7. T   8. T   9. F   10. F
11. b   12. a   13. b   14. b   15. c   16. d   17. a   18. b   19. c   20. a
21. middle-latitude climates   22. elevation   23. polar climates   24. a monsoon
25. cools   26. equator   27. temperature   28. precipitation
29. doldrums   30. tropical

31. The average daily temperature is 21°C. The temperature range is 18°C.

32. In the winter, the land loses heat more quickly than the water. A high-pressure system develops over the land and the cool air flows away from the land toward the ocean. Thus the wind moves seaward. In summer, low pressure develops over the land. Warm air over the land rises and is replaced by cool air flowing in from over the ocean. Thus the wind moves landward.

33. The steady westerly winds are responsible for both effects. They blow toward northwestern Europe carrying warmth from the Gulf Stream, but blow away from the eastern United States.

34. The tropical savanna climate results when an area has both a rain forest climate and a tropical desert climate at different times. During different seasons the precipitation belts shift toward different poles, producing very wet summers and very dry winters in these regions.

35. The Sierra Nevada mountains influence the temperature and moisture content of the passing air masses. As the mild air rises from the Los Angeles area up the west side of the mountains, it cools and loses most of any moisture it contains through precipitation. As the air descends into Death Valley on the east side, it is warmed. The air moving into Death Valley, therefore, becomes very warm and dry.

Guide to Tested Objectives

Objective 1: questions 11(A), 26(A), 27(A), 29(A), 31(A); Objective 2: questions 2(A), 9(B), 10(A), 12(A), 24(A), 32(B), 33(A); Objective 3: questions 1(A), 13(A), 25(A), 35(B); Objective 4: questions 3(A), 6(A), 21(A), 28(A), 30(A), 34(A); Objective 5: questions 4(A), 15(A), 17(A), 23(A); Objective 6: questions 5(A), 14(A), 18(A), 19(A), 20(B); Objective 7: questions 7(A), 8(A), 16(B), 22(B)

M O D E R N   E A R T H   S C I E N C E

Answers to Unit 7 Test

1. F	2. T	3. F	4. T	5. T	6. F	7. F	8. T	9. T	10. F
11. T	12. T	13. T	14. T	15. T	16. b	17. c	18. c	19. b	20. d
21. d	22. b	23. c	24. c	25. d	26. a	27. c	28. d	29. c	30. a

31. cP

32. west to east   33. radiation fog   34. specific heat   35. waterspout

36. the ionosphere   37. dew point   38. precipitation   39. a barometer

40. the atmosphere   41. rain forests   42. radiosonde   43. tundra

44. specific humidity

45. Polar fronts generally circle the earth between 40° and 60° latitude in each hemisphere. They move closer to the equator in the winter and back toward the poles in the summer.

46. Chinooks are warm, dry winds that flow down the eastern slopes of the Rocky Mountains. These winds can raise the temperature very rapidly.

47. Cloud seeding involves the release of silver-iodide vapor or powdered dry ice into supercooled clouds in an attempt to cause or increase precipitation. In supercooled clouds, precipitation forms when water evaporates from supercooled droplets and condenses onto ice crystals. However, there are few ice crystals because there are few freezing nuclei. Cloud seeding adds freezing nuclei to the supercooled clouds.

48. As some of the air is heated, it becomes less dense and tends to rise. Nearby cooler air, which is more dense, tends to sink. As it sinks, the cooler air pushes the warm air up. The cold air is, in turn, warmed and then rises. The continuous cycle helps warm the earth's surface.

49. Convective currents within cumulonimbus clouds carry raindrops upward to high altitudes, where they freeze. As the frozen raindrops fall, they accumulate additional layers of liquid water on their surfaces. These layers freeze when the hail is carried aloft again or encounters other near-freezing moist air masses. Hailstones may consist of many layers of ice.

50. Short-wavelength rays, such as blue light rays, are more easily scattered than long-wavelength rays are. The sun appears red when it is low in the sky because more blue light rays are scattered by the atmosphere. Thus more of the longer-wavelength red light rays reach the earth's surface, giving the setting sun its red color.

Guide to Tested Objectives

Chapter 23—Objective 1: questions 30(A), 40(A); Objective 2: question 39(A); Objective 3: question 36(A); Objective 4: question 5(A); Objective 5: question 20(A); Objective 6: questions 13(B), 22(A), 50(B); Objective 7: questions 12(A), 48(B); Objective 8: question 21(A); Objective 9: question 6(A). **Chapter 24**—Objective 1: question 3(A); Objective 2: questions 16(B), 42(B), 44(A); Objective 3: question 37(A); Objective 4: question 4(B); Objective 5: question 17(A); Objective 6: question 33(A); Objective 7: questions 15(A), 49(A)-feature; Objective 8: question 25(A); Objective 9: question 47(B); Objective 10: question 26(A). **Chapter 25**—Objective 1: question 9(B); Objective 2: questions 23(B), 31(B); Objective 3: questions 7(A), 29(B); Objective 4: questions 24(A), 32(A), 45(B); Objective 5: question 35(A); Objective 6: question 8(A); Objective 8: question 14(A); Objective 9: question 38(A). **Chapter 26**—Objective 1: question 11(A); Objective 2: questions 1(A), 34(A); Objective 3: questions 18(B), 46(A); Objective 4: questions 19(B), 41(A); Objective 5: questions 2(A), 43(A); Objective 6: questions 27(A), 28(A); Objective 7: question 10(A).

HRW material copyrighted under notice appearing earlier in this work.

215

M O D E R N   E A R T H   S C I E N C E

Answers to Chapter 27 Test

1. T	2. T	3. F	4. F	5. T	6. F	7. F	8. F	9. T	10. F
11. c	12. b	13. d	14. a	15. d	16. b	17. a	18. c	19. d	20. b

21. constellations   22. quasars   23. nuclear fusion   24. globular clusters
25. parallax   26. Local Group   27. main-sequence phase   28. absolute magnitude
29. circumpolar   30. light-years

31. Astronomers theorize that a black hole can be located by observing its effect on a companion star. Just before matter from a companion star is pulled into a black hole, it gives off X rays. Astronomers try to locate the black holes by detecting these X rays from the earth.

32. The circular trails left by most stars suggest that they are moving in a circle around Polaris, the North Star. This circular pattern actually results from the rotation of the earth on its axis. Polaris is located almost directly above the North Pole and does not appear to move.

33. According to the big bang theory, all of the matter and energy in the universe was once concentrated in an extremely small volume. About 17 billion years ago, the big bang occurred, sending this matter and energy out in all directions. As matter and energy moved outward with the expanding universe, gravity began to affect the matter, causing it to condense and form galaxies.

34. A Cepheid variable is a type of star that brightens and fades in a regular pattern caused by the swelling and shrinking of the star. Astronomers measure the cycle of brightness changes and estimate the star's true brightness. Then they compare the Cepheid's apparent brightness with its true brightness and calculate the distance to the galaxy where the Cepheid is located.

35. The star has most likely become a giant. Giants are ten times bigger than the sun. In the giant stage, hydrogen is no longer a source of fuel, and the star's core contracts under the force of its own gravity. This contraction causes increased pressure in the core, causing the helium atoms there to fuse into carbon atoms.

Guide to Tested Objectives

Objective **1**: questions 3(A), 8(A), 24(A); Objective **2**: questions 11(A), 14(A), 29(A), 32(B); Objective **3**: questions 25(A), 30(A), 34(B); Objective **4**: questions 15(B), 19(B), 20(B); Objective **5**: questions 2(A), 18(A), 23(A); Objective **6**: questions 1(A), 27(A); Objective **7**: questions 12(A), 16(A), 31(B), 35(B); Objective **8**: questions 7(A), 13(B), 21(A); Objective **9**: questions 4(A), 6(A), 9(A), 10(A)-feature, 17(A), 26(A); Objective **10**: questions 5(A), 22(A), 28(A), 33(A)

Answers to Chapter 28 Test

1. F	2. T	3. T	4. F	5. T	6. F	7. T	8. F	9. T	10. T
11. b	12. a	13. a	14. d	15. b	16. b	17. a	18. c	19. d	20. c

21. corona   22. radiative zone   23. core   24. magnetic storms
25. photosphere   26. condensed   27. iron and nickel   28. moons
29. sunspots   30. auroras

31. Magnetic fields slow down activity in the sun's convective zone, so less heat is transferred from the core to the photosphere. Areas of the photosphere near these magnetic fields that are cooler and appear darker than surrounding areas are called sunspots.

32. The solar nebula contracted and became denser due to shock waves from a nearby supernova or some other force. The center became denser and hotter due to heat from collisions and pressure caused by the force of gravity. When the temperature in the nebula's center became great enough, hydrogen fusion occurred and the sun formed.

33. In the convective zone, as hot gas atoms move outward and expand, they radiate and lose heat. These cooler gases become denser and sink to the bottom of the convective zone, where they are heated by energy from the radiative zone and rise again.

34. Because the sun is a ball of hot gases rather than a solid sphere, the parts of the sun rotate at different speeds. Places closest to the sun's equator rotate faster than points near the poles.

35. If the earth were more massive, its gravitational pull would be greater, and it would have been able to retain most of the hydrogen that originally was present but escaped into space. Also, if no green plants existed, then there would be no photosynthesis. Thus the amount of hydrogen would increase and the amount of oxygen would decrease.

Guide to Tested Objectives

Objective **1**: questions 3(A), 6(A), 14(A), 20(B), 23(A); Objective **2**: questions 8(A), 15(A), 22(A), 33(B); Objective **3**: questions 19(A), 25(A); Objective **4**: questions 11(A), 29(A), 31(B), 34(A); Objective **5**: questions 2(A), 12(A), 24(A); Objective **6**: questions 1(A), 18(A), 30(A); Objective **7**: questions 4(B), 9(B), 17(A), 21(A), 32(A); Objective **8**: questions 5(A), 10(A), 13(A), 28(A); Objective **9**: questions 7(A), 16(A), 26(A), 27(B), 35(B)

M O D E R N   E A R T H   S C I E N C E

Answers to Chapter 29 Test

1. F	2. F	3. T	4. T	5. F	6. T	7. T	8. F	9. F	10. T
11. d	12. d	13. c	14. a	15. d	16. a	17. b	18. c	19. d	20. c

21. Galileo
22. an ellipse
23. impact craters
24. Jovian planets
25. Trojan asteroids
26. meteor shower
27. Ptolemy
28. a short-period comet
29. *Voyager 2*
30. a meteorite

31. Since planetary orbits are ellipses, the distance from the center of a planet to the center of the sun is not always the same. Because a planet moves fastest when it is closest to the sun, an imaginary line along the orbit's moving radius creates, over a given period of time, a wide triangular sector. When the planet is farther from the sun, the sector created in the same time period is longer and more narrow. Both sectors, however, will have equal areas.

32. Since Mercury is so close to the sun, solar heat causes the gas molecules near the surface of the planet to move rapidly. Because the planet is so small, its gravitational pull is too weak to hold enough rapidly moving gas molecules to form a dense atmosphere.

33. The temperature is warm enough for water to exist as a liquid. Mercury and Venus are too close to the sun to retain liquid water. The outer planets are too far from the sun, and most of their water is in the form of ice.

34. Most planets rotate with their axes perpendicular to their orbital planes as they revolve around the sun. Uranus, however, appears to rotate like a rolling ball. The axis of Uranus is almost horizontal to the plane of its orbit.

35. A moving object will continue to move in a straight line unless something causes it to change direction or stop. A stationary object will remain at rest until an outside force acts on it.

Guide to Tested Objectives

Objective 1: questions 1(A), 2(A), 21(A), 27(A); Objective 2: questions 3(A), 11(B), 12(A), 22(A), 31(B); Objective 3: questions 4(A), 5(A), 15(A), 23(A), 32(B), 35(B); Objective 4: questions 6(A), 13(A), 16(A), 33(A); Objective 5: questions 7(A), 18(A), 20(B), 24(A); Objective 6: questions 8(A), 17(B), 19(A), 29(B)-feature, 34(A); Objective 7: questions 9(A), 14(A), 25(A), 28(A); Objective 8: questions 10(A), 26(A), 30(A)

Answers to Chapter 30 Test

1. F	2. T	3. F	4. F	5. T	6. F	7. T	8. T	9. T	10. F
11. d	12. a	13. a	14. a	15. d	16. b	17. a	18. b	19. a	20. c

21. 384,000 km
22. apogee
23. the mantle
24. new moon phase
25. 27.3 days
26. umbra
27. earthshine
28. one month (29.5 days)
29. Mars
30. Neptune

31. The moon formed when a Mars-sized body struck the earth causing an ejection of fragments into orbit around the earth. These fragments eventually joined to form the moon. Most of the ejected materials were from the silicate-rich mantles of the earth and the colliding body, while dense, metallic core materials remained in the earth.

32. During an annular solar eclipse, the moon comes between the earth and sun but its umbra does not reach the earth. The sun is not completely blocked out and appears as a bright ring around the moon.

33. One theory hypothesizes that Saturn's rings formed from the breakup of a satellite of Saturn; the other theory hypothesizes that the rings are made of material that was unable to condense into a moon.

34. The surfaces differ because the earth has an atmosphere and the erosional agents of water, wind, and ice that alter the earth's surface, while the moon's surface remains largely the same as it was 3 billion years ago.

35. An individual on the moon weighs much less than he or she does on the earth. Since the moon has less mass than the earth does, its gravity is weaker. An individual's weight is determined by the force of gravity acting on the individual. The moon's surface gravity is about one-sixth the surface gravity of the earth, so an individual on the moon would weigh about one-sixth of what he or she weighs on the earth.

Guide to Tested Objectives

Objective 1: questions 1(A), 11(A), 18(A), 34(A), 35(B); Objective 2: questions 2(A), 10(B)-feature, 23(A); Objective 3: questions 3(A), 31(B); Objective 4: questions 4(A), 13(A), 21(B), 22(A), 25(A); Objective 5: questions 5(A), 12(B), 19(B), 20(B), 26(A), 32(A); Objective 6: questions 6(A), 14(A), 24(A), 27(A); Objective 7: questions 7(A), 15(A), 28(A); Objective 8: questions 8(A), 16(A), 29(A); Objective 9: questions 9(B), 17(A), 30(A), 33(B)

M O D E R N   E A R T H   S C I E N C E

Answers to Unit 8 Test

1. F	2. T	3. F	4. T	5. T	6. T	7. T	8. T	9. F	10. F
11. F	12. F	13. T	14. T	15. T	16. a	17. d	18. a	19. c	20. b
21. d	22. b	23. a	24. b	25. c	26. c	27. d	28. a	29. d	30. a
31. b									

32. geocentric
33. erupting volcanoes
34. Saturn
35. nebula
36. energy
37. quasars
38. Polaris (the North Star)
39. Mars
40. highlands
41. black holes
42. electromagnetic waves
43. prominences
44. stony-iron

45. It stated that the sun and the planets condensed out of the same spinning nebula. It also stated that the entire solar system formed at approximately the same time.

46. The core or nucleus is made up of rock or metals and ice. The coma surrounding the nucleus is a cloud of gas and dust. The tail is the gas and dust that stream out from the head.

47. Parallax is used to determine the distance to a nearby star by observing it from slightly different angles. During a six-month period, a nearby star will appear to shift slightly relative to stars that are farther away. The closer a nearby star is, the greater will be the amount of shift. From the amount of shift, astronomers can calculate the distance to any star within 1,000 light-years.

48. Astronomers think novas occur in white dwarfs that revolve around a giant or main-sequence star. The white dwarf has a greater surface gravitational attraction because it is denser, and gases from the companion star accumulate on the white dwarf. This causes the pressure in the white dwarf to increase until it explodes, creating a nova.

49. The moon rotates on its axis very slowly, completing a rotation only once during each orbit around the earth. Because the moon rotates at the same rate that it revolves around the earth, the same side of the moon always faces the earth. From an earth-bound viewpoint, we can only ever see the near side of the moon.

50. First, the earth retained much of the heat produced when it collided with planetesimals. Second, the increasing weight of the outer layers compressed the inner layers and generated heat. Third, high-energy particles emitted from radioactive materials were absorbed by rocks. Some of the energy of motion of the particles was converted to heat.

Guide to Tested Objectives

Chapter 27—Objective **1**: question 29(A); Objective **2**: question 38(A); Objective **3**: questions 10(A), 15(A); Objective **4**: question 6(A); Objective **5**: question 47(B)-feature; Objective **6**: question 30(A); Objective **7**: questions 28(A), 41(A), 48(B); Objective **8**: question 22(A); Objective **9**: question 11(B); Objective **10**: question 37(B)-feature. **Chapter 28**—Objective **1**: questions 4(A), 36(A); Objective **2**: question 42(A); Objective **3**: question 26(A); Objective **4**: question 31(B); Objective **5**: question 43(A); Objective **6**: question 5(A); Objective **7**: questions 35(A), 45(B); Objective **8**: question 18(A); Objective **9**: questions 12(A), 19(A), 50(B). **Chapter 29**—Objective **1**: questions 2(A), 32(A); Objective **2**: questions 17(A), 21(B); Objective **3**: question 33(A); Objective **4**: question 25(A); Objective **5**: questions 3(A), 20(B); Objective **6**: question 13(A); Objective **7**: questions 14(A), 46(B); Objective **8**: question 44(A). **Chapter 30**—Objective **1**: questions 1(A), 40(B); Objective **2**: question 7(A); Objective **3**: question 8(A); Objective **4**: question 49(B); Objective **5**: question 23(A); Objective **6**: question 27(A); Objective **7**: questions 16(A), 24(B); Objective **8**: question 39(A); Objective **9**: questions 9(A), 34(A).

M O D E R N   E A R T H   S C I E N C E

Answers to Final Test

1. T	2. F	3. T	4. T	5. T	6. F	7. T	8. T	9. F	10. F
11. T	12. F	13. T	14. F	15. F	16. T	17. F	18. F	19. b	20. a
21. a	22. b	23. d	24. d	25. c	26. c	27. b	28. a	29. d	30. a
31. d	32. d	33. b	34. c	35. a	36. d	37. a	38. c	39. c	40. b
41. a	42. b	43. a	44. d	45. b	46. a	47. c	48. c	49. a	50. d
51. c	52. a	53. c	54. a	55. b	56. b	57. c	58. a	59. b	60. d
61. a	62. d	63. c	64. b	65. c	66. b	67. c	68. a	69. b	70. c
71. a									

72. longitude
73. scale
74. topographic map
75. continental drift
76. an element
77. law of superposition
78. relative age
79. unconformities
80. epochs
81. invertebrates
82. ooze
83. continental shelf
84. sonar
85. continental air mass
86. Coriolis effect
87. absolute magnitude
88. atmosphere
89. hydrogen
90. sun
91. nucleus
92. calcium carbonate
93. glacial periods
94. kettles
95. mantle

96. Large lakes can cause an increase in precipitation on the shore downwind. The eastern shore of Lake Michigan generally has more moderate temperatures, more cloudiness, and higher precipitation than the western shore does.

97. The outer layer or crust is 60-100 km thick. Below the crust is a dense mantle of rocks rich in silica, magnesium, and iron. The mantle extends to a depth of 1,000 km. At the center may be a small iron core. The lower portion of the mantle and the core may be slightly molten and the core may even be liquid.

98. The Canadian Shield is a large area in eastern Canada and parts of the northeastern United States where very old Precambrian rocks are exposed. It is the exposed portion of the craton around which the modern continent of North America has been built up.

99. A mixture is material that contains two or more substances that are not chemically combined. The substances in a mixture keep their individual properties. Therefore, unlike a compound, a mixture can be separated into its parts by physical means.

100. Coral reefs are formed by corals, which are small animals that live in warm, shallow waters. A coral extracts calcium carbonate from ocean water and uses it to build a hard outer skeleton. Corals attach to one another to form large colonies. New corals grow on top of dead ones, forming a coral reef, which is a ridge made up of millions of coral skeletons.

Guide to Tested Objectives

Chapter 1—Objective 2: question 1(A); Objective 3: question 19(A); Objective 6: question 2(A); Objective 7: question 20(A).
Chapter 2—Objective 1: questions 21(A), 95(A); Objective 6: question 22(A); Objective 8: question 23(A).
Chapter 3—Objective 1: question 72(A); Objective 5: question 73(A); Objective 6: question 74(A). **Chapter 4**—Objective 1: question 75(A); Objective 5: question 24(A); Objective 7: question 25(A). **Chapter 5**—Objective 2: question 26(A); Objective 3: question 27(A); Objective 5: question 28(A). **Chapter 6**—Objective 1: questions 29(A), 30(A); Objective 5: question 31(B). **Chapter 7**—Objective 2: question 3(A); Objective 3: question 4(A); Objective 6: question 32(A).
Chapter 8—Objective 1: question 76(A); Objective 3: question 6(A); Objective 7: question 5(A); Objective 9: question 99(A).
Chapter 9—Objective 1: question 33(A); Objective 3: question 35(B); Objective 5: question 34(A). **Chapter 10**—Objective 1: question 36(A); Objective 2: question 39(A); Objective 6: question 38(B); Objective 8: question 37(B).
Chapter 11—Objective 1: question 41(A); Objective 3: question 40(A); Objective 8: question 42(A); Objective 11: question 43(A). **Chapter 12**—Objective 2: question 45(A); Objective 7: question 92(B); Objective 9: question 44(A).
Chapter 13—Objective 1: questions 70(B), 71(A). **Chapter 14**—Objective 1: question 7(B); Objective 5: question 8(B).
Chapter 15—Objective 4: question 47(A); Objective 5: questions 46(B), 94(A); Objective 7: question 93(A).
Chapter 16—Objective 1: question 66(A); Objective 3: question 67(A); Objective 6: question 68(A); Objective 8: question 100(B). **Chapter 17**—Objective 2: question 77(A); Objective 3: question 79(A); Objective 7: question 48(A); Objective 10: question 78(A). **Chapter 18**—Objective 1: question 49(A); Objective 3: question 81(A); Objective 6: questions 50(B), 80(B). **Chapter 19**—Objective 1: question 9(A); Objective 3: question 98(B); Objective 4: question 10(B).
Chapter 20—Objective 2: question 84(A); Objective 3: question 83(B); Objective 6: question 82(A). **Chapter 21**—Objective 2: question 52(A); Objective 5: question 51(A). **Chapter 22**—Objective 4: question 11(B); Objective 5: question 12(A).
Chapter 23—Objective 3: question 54(A); Objective 6: question 53(B); Objective 8: question 55(A). **Chapter 24**—Objective 2: question 56(A); Objective 5: question 57(A); Objective 7: question 58(B). **Chapter 25**—Objective 1: questions 59(A), 86(A); Objective 2: question 85(B); Objective 6: question 60(A); Objective 8: question 61(A). **Chapter 26**—Objective 1: question 13(A); Objective 6: question 14(A); Objective 7: question 96(B). **Chapter 27**—Objective 2: question 17(A); Objective 4: question 87(A); Objective 9: question 18(A). **Chapter 28**—Objective 1: question 89(B); Objective 3: questions 63(A), 88(A); Objective 4: question 62(A); Objective 8: question 69(A). **Chapter 29**—Objective 1: question 64(A); Objective 2: question 90(A); Objective 3: question 65(A); Objective 7: question 91(A). **Chapter 30**—Objective 2: question 97(B); Objective 3: question 15(B); Objective 5: question 16(A).